55: A Decade of Experience

Transportation Research Board
Special Report 204

National Research Council

Washington, D.C. 1984

Transportation Research Board Special Report 204
Price $14.00
Edited for TRB by Edythe Traylor Crump

mode
1 highway transportation

subject areas
17 energy and environment
51 transportation safety
54 operations and traffic control

Transportation Research Board publications are available by ordering directly from TRB. They may also be obtained on a regular basis through organizational or individual affiliation with TRB; affiliates or library subscribers are eligible for substantial discounts. For further information, write to the Transportation Research Board, National Research Council, 2101 Constitution Avenue, N.W., Washington, D.C. 20418.

NOTICE
The project that is the subject of this report was approved by the Governing Board of the National Research Council, whose members are drawn from the councils of the National Academy of Sciences, the National Academy of Engineering, and the Institute of Medicine. The members of the committee responsible for the report were chosen for their special competence and with regard for appropriate balance.

This report has been reviewed by a group other than the authors according to procedures approved by a Report Review Committee consisting of members of the National Academy of Sciences, the National Academy of Engineering, and the Institute of Medicine.

The study was sponsored by the National Highway Traffic Safety Administration of the U.S. Department of Transportation.

Library of Congress Cataloging in Publication Data
National Research Council. Transportation Research Board.
55, a decade of experience.
(Special report; 204)
1. Speed limits—United States—Cost effectiveness. I. National Research Council (U.S.). Transportation Research Board. II. Title: Fifty-five, a decade of experience. III. Series: Special report (National Research Council (U.S.). Transportation Research Board); 204.
HE5620.S6A154 1984   388.3′144′0973   84-16547
ISBN 0-309-03664-X   ISSN 0360-859X

Printed in the United States of America

COMMITTEE FOR THE STUDY OF THE BENEFITS AND COSTS OF THE 55 MPH NATIONAL MAXIMUM SPEED LIMIT

ALAN A. ALTSHULER, New York University, *Chairman*
HARRY T. ADAIR, California Highway Patrol, Sacramento
JOHN R. BORCHERT, University of Minnesota, Minneapolis
JOSEPH M. CLAPP, Roadway Express, Incorporated, Akron, Ohio
W. DALE COMPTON, Ford Motor Company, Dearborn, Michigan
R ADAMS COWLEY, University of Maryland, Baltimore
BENJAMIN O. DAVIS, Lt. General USAF (retired), Arlington, Virginia
JOHN J. FEARNSIDES, The MITRE Corporation, McLean, Virginia
PAUL H. FOWLER, Automobile Club of Southern California, Los Angeles
TREVOR O. JONES, TRW, Incorporated, Solon, Ohio
CHARLES A. LAVE, University of California, Irvine
DARRELL V MANNING, Idaho Transportation Department, Boise
H. LAURENCE ROSS, University of New Mexico, Albuquerque
PETER J. SAFAR, University of Pittsburgh, Pittsburgh, Pennsylvania
WAYNE W. SORENSON, State Farm Insurance Companies, Bloomington, Illinois
EVELYN I. TEEGEN, National Highway Safety Advisory Committee, Edina, Minnesota
PATRICIA F. WALLER, University of North Carolina, Chapel Hill
JOHN J. WIORKOWSKI, University of Texas, Dallas
MARCUS L. YANCEY, Texas State Department of Highways and Public Transportation, Austin

Liaison Representatives

FRANK G. EPHRAIM, National Highway Traffic Safety Administration, Washington, D.C.
KAREN L. PHILLIPS, Committee on Commerce, Science and Transportation, U.S. Senate, Washington, D.C.
CARYLL F. RINEHART, U.S. House of Representatives, Washington, D.C.
JAMES K. WILLIAMS, Transportation Research Board, Washington, D.C.

Transportation Research Board Staff

DAMIAN J. KULASH, Assistant Director for Special Projects
FRANCIS P. MULVEY, Program Manager
STEPHEN R. GODWIN, Program Manager

Preface

In 1982 Congress expressed concern that eroding compliance with the 55 mph national maximum speed limit threatened the safety of America's highways. Several states had weakened their penalties for violations of the 55 mph speed limit, and some members of Congress became concerned that this would encourage faster driving and impair highway safety. In the Surface Transportation Assistance Act of 1982, Congress requested that the National Academy of Sciences "investigate (1) the benefits, both human and economic, of lowered speeds due to the enactment of the 55 mile per hour National Maximum Speed Limit, with particular attention to savings to the taxpayers, and (2) whether the laws of each State constitute a substantial deterrent to violations of the maximum speed limit on public highways within such State." The legislation also requested "a report to the Secretary [of Transportation] and the Congress . . . on the results of the study together with its recommendations."

Under support from the National Highway Traffic Safety Administration, the National Research Council assembled a committee of 19 individuals with expertise in the various disciplines needed to evaluate the benefits and costs of the 55 mph speed limit and to assess the effectiveness of state laws in inducing compliance. The committee included individuals with expertise in statistics, highway safety, economics, trucking, insurance, transportation administration, law enforcement, medical services, traffic engineering, law, sociology, motor vehicle design, and public policy.

The results of this study are contained in this report. To communicate the nature of the trade-offs involved, the committee made numerical estimates of the benefits and costs of the 55 mph speed limit, even though measurement of most effects is imprecise. In determining the effects on safety, analysis is confounded by the difficulty of isolating the effects of the speed limit from other causes of the improved safety record. Indirect estimating techniques must be relied on, and assumptions must be made in the process. The committee believes that in spite of the difficulties some rough estimates can be offered, based on the plausibility of the techniques employed and the similarity of findings that emanate from different data and different statistical methods. Nevertheless, an exact determination of a specific number of lives saved by the 55 mph speed limit is not possible. Data on the effects on serious injuries are particularly sketchy, and any estimate of the effect of the speed limit on injuries is essentially an educated guess.

The collection or analysis of additional data from the 1970s does not promise to alleviate these shortcomings. Missing data on speeds, fatalities, and injuries during the early to mid-1970s cannot be reconstructed at this stage. Although innovative analytical techniques might yield new insights from the existing data, the committee believes that it can best contribute by examining and reconciling the results of numerous previous analyses of the effect of speed limits that have used national data, the experience of specific states or toll roads, and data collected in other nations.

The predominant views of the committee are presented in this report. However, on any particular finding or recommendation, individual members of the committee might disagree with the predominant view. To simplify the presentation, such disagreements have not been noted unless the committee was divided, that is, if more than two-thirds of the committee members could not agree on a single position. When disagreements precluded a consensus finding, the nature of the debate is revealed in the discussion of the issues.

For example, the committee discussed in great detail whether there were ways to relax the speed limit on roads where the life saving benefits were relatively small and the time costs were relatively large while enacting other safety actions so that a net improvement in safety would result. Different committee members advanced diverse ways by which such a concept might be enacted. Ultimately, however, most members concluded that this idea, despite its conceptual appeal, would not be technically and administratively feasible.

Presented in this report are the collective findings and arguments of a dedicated study committee, Transportation Research Board staff, and many interested individuals who provided valuable data, insights, and suggestions for this study. The committee was chaired by Dr. Alan

Altshuler, Dean of the Graduate School of Public Administration at New York University; his guidance and direction are greatly appreciated. The study was performed under the supervision of Dr. Damian J. Kulash, Assistant Director, Dr. Francis P. Mulvey, and Stephen R. Godwin of the Special Projects Division of TRB. Dr. Lindsay I. Griffin, III, consultant, greatly assisted the staff on many technical issues.

Nancy A. Ackerman, TRB Publications Manager, and Edythe T. Crump, Associate Editor, deserve special appreciation for editing the final report, as does Margaret M. Sheriff for typing the many drafts and the final copy of this study.

Contents

Executive Summary

To assist the Congress in setting policy on the 55 mph national maximum speed limit, the conclusions of the Committee for the Study of Benefits and Costs of the 55 mph National Maximum Speed Limit are summarized as follows:

- *Findings* of the committee about the effect of the 55 mph speed limit on safety, energy, travel time, and taxpayer costs;
- *Recommendations* of the committee for continuation of the 55 mph speed limit and the mechanism by which state compliance is measured; and
- *Unresolved Issues* that surround the appropriate speed limit for selected roads.

FINDINGS

Enacted during the Arab oil embargo to conserve fuel, the 55 mph speed limit became effective in all states by March 1974. Motorists slowed down on all major highway systems. Total travel declined in 1974 for the first time since 1946. Accompanying these reductions in speed and travel, 9,100 fewer persons died in motor vehicle accidents. The number of highway fatalities declined from 55,511 in 1973 to 46,402 in 1974. Such a sudden annual decline in highway fatalities was unprecedented outside of wartime, and as the fuel shortage receded, safety soon became the paramount issue surrounding the 55 mph speed limit.

Consequences of the 55 mph Speed Limit in 1974

Not all of the 9,100 fewer fatalities that occurred on the nation's highways in 1974 can be attributed to the 55 mph speed limit, however. Part of the reduction in fatalities in 1974 can be attributed to economic conditions and fuel shortages that curtailed the amount of driving. Improvements in safety due to improved vehicle characteristics, safer highways, and increased emergency medical services also contributed to the reduced number of highway deaths.

The magnitude of the decline in fatalities, however, appears to be far larger than can be explained by reduced travel or improved vehicles, highways, and medical services. Although travel declined, the reductions in highway fatalities, and fatalities per mile traveled, declined far more. For example, toll roads—highways with the most accurate travel records—demonstrated a much more pronounced decline in fatalities per mile traveled than could be expected based on the decline in travel. The nationwide decline in the number of fatalities per mile traveled (15.3 percent) was the sharpest decline in this rate recorded in peacetime.

In response to improvements in safety, many states examined their speed and fatality records and concluded that much of the safety improvement in 1974 resulted from the enactment of the 55 mph speed limit. In numerous studies of state experience this conclusion was reached despite substantial variations in highway design standards and differences in the availability of emergency medical services. Although interpretation of safety trends is confounded by gaps and inconsistencies in the data collected on travel, vehicle speeds, and other factors that influence highway safety, independent statistical analyses of national data using a variety of approaches arrive at a similar conclusion: some combination of factors significantly shifted the trend in highway fatalities in 1974. Further, experience in numerous foreign countries indicates that reductions in speed limits are virtually always accompanied by reductions in injuries and/or fatalities. Together these findings indicate that the slower speeds and more uniform pace of travel due to the 55 mph speed limit accounted for 3,000 to 5,000 fewer highway fatalities in 1974.

Current Consequences of the 55 mph Speed Limit

Safety

The reduction in fatalities evidenced in 1974 does not appear to have been simply a temporary, passing phenomenon. Although the average speed driven by motorists on rural Interstates has crept upward from 57.6 mph in 1974 to 59.1 mph in 1983, it is still far below the average of 65 mph that prevailed in 1973. More important, the distribution of highway speeds continues to be narrower than before the 55 mph speed limit became effective, and this reduced variance enhances highway safety. The physical relationships that govern motorist reaction times and crash impact have not fundamentally changed, and the sustained restraint of highway speeds during the last decade continues to affect highway safety through such relationships. Nevertheless, the reduced percentage of teenage drivers; increased availability of improved emergency medical services; new energy-absorbing barriers at dangerous locations on highways; and numerous other changes affecting highways, vehicles, drivers, and medical treatment have resulted in a net reduction in the fatality rate of 23 percent between 1974 and 1983, even though speeds have risen slightly.

Most highway travel today occurs in vehicles that were not on the road in 1974, and many new safety features are incorporated in the cars now in use. Because of improvements to roads and vehicles, the effect of the 55 mph speed limit has probably declined somewhat. For example, if a bridge-abutment was protected with an energy-absorbing barrier in 1980, a high-speed accident that occurred at such a location today may not be fatal to the driver although it might have been fatal a decade ago in the absence of the energy-absorbing barrier. Adjusting for the reduction in the fatality rate during the last decade as well as the increase in travel and average speeds during that period, about 2,000 to 4,000 lives were saved in 1983 as a result of the 55 mph speed limit. In addition, about 2,500 to 4,500 fewer serious, severe, and critical injuries and 34,000 to 61,000 fewer minor and moderate injuries occurred in 1983 because of the slower speeds and more uniform pace of travel brought about by the 55 mph speed limit.

Energy

In addition to the safety benefits, the 55 mph speed limit saves energy. The speed limit was reduced in 1974 as an energy conservation measure, and currently some 167,000 barrels of petroleum per day are saved because of reduced speeds. At current crude oil prices this represents a saving of $2 billion annually. Although this is less than 2 percent of the nation's highway-fuel consumption, this difference could be important in the event of another disruption in oil supply. For example, the annual fuel savings created by the 55 mph speed limit represent about 10 percent of the shortfall experienced during the Arab oil embargo.

Taxpayer Costs

Several government programs provide benefits that may be affected by highway fatalities, disabilities, and injuries. The chief programs are Medicare, Medicaid, and Old Age Survivors and Disability Insurance. Taxpayers pay about $65 million less per year for such programs because of the 55 mph speed limit; this corresponds to a reduction in total federal spending of about 0.01 percent.

Travel Time

The chief cost of the 55 mph speed limit, additional travel time, amounts to about 1 billion hours per year. This means that each highway user spends an average of about 7 hours more per year on the road than if driving had continued at 1973 speeds. On average, 350,000 hours of additional travel time are expended by motorists each year for the saving of one life plus the avoidance of one serious, severe, or critical injury.

Although no meaningful comparison can be made between a year of additional driving time and an additional year of life, these figures imply that the effect of the 55 mph speed limit is to gain approximately one year of life for the expenditure of one year of driving time. Further, much of the additional driving time that comprises this total is generated in small increments of 1 or 2 minutes per trip, and the appropriate value of such time is arguable. Not all motorists necessarily find the added travel time burdensome. Some motorists, particularly the elderly, prefer to drive in moderately paced traffic streams.

Despite the increased travel time caused by slower speeds, the majority of Americans support the 55 mph speed limit. The most recent national poll indicates that 76 percent of the public supports the law, and also indicates majority support across all regions:

Region	*Favor*	*Oppose*	*No Opinion*
Northeast	81	17	2
Midwest	75	24	1
South	70	28	2
West	72	25	3

The additional travel time caused by the 55 mph speed limit is most frustrating to long distance truck drivers, intercity bus drivers, traveling salesmen, repairmen, and others who expend far more time driving on high speed highways than the national average. The economic cost of additional travel time is partly offset by reduced fuel, maintenance, and insurance costs, and as a result, most large trucking firms support the 55 mph speed limit. Residents in rural areas of the United States drive more than the average American on highways posted at 55 mph, and the above average cost in additional travel time for residents of predominantly rural states may explain the somewhat lower level of support for the law outside of the Northeast.

Recent Trends in Motorist Compliance and Pressures for Change

In contrast to the overwhelming support for the 55 mph speed limit, compliance with the law has declined markedly in recent years. Speed data collected from automated devices embedded in pavements indicate that in 1983 more than one-half of all drivers on roads posted at 55 mph exceeded the speed limit in 37 states, up from 30 states in 1982. Under the provisions of current law, this means that more states are nearing the point of technical noncompliance, and hence will be threatened with the loss of some federal-aid highway funding. Although the Gallup Poll indicates that 76 percent of the public supported the speed limit in 1982, this compares with 81 percent 2 years earlier. Legislatures in several states have passed measures that deemphasize the importance of the 55 mph speed limit.

These developments are alarming to some proponents of highway safety, who interpret them as a gradual nullification of the 55 mph speed limit that will lead to forfeiture of the safety and energy contributions this law has made. Policymakers will be challenged to enact a strategy that preserves the safety and energy gains of the 55 mph speed limit, but also accommodates trends in highway speeds, public opinion polls, and the doubts expressed by state legislatures.

RECOMMENDATIONS

National Maximum Speed Limit

Recognizing the safety and energy implications and the time consequences of the 55 mph speed limit during the past decade, the committee believes that the fundamental shift in driver behavior patterns that began in 1974 has been an improvement. The committee recommends that the 55 mph speed limit be retained on most of the highways now posted at 55 mph, although members disagreed on whether it should be retained on all highways or whether some carefully controlled exceptions should be permitted, as discussed later.

TABLE 1 Major Effects of the 55 mph Speed Limit

Impact Area	Total for 1983	Effect of Speed Limit	Percentage
Highway fatalities	44,300	(−) 2,000 to 4,000	(−) 4 to 8
Transportation			
Petroleum consumption (000s of barrels per day)	9,200	(−) 167	(−) 1.8
Motor vehicle accident			
Costs borne by taxpayers ($ millions)	2,500	(−) 65	(−) 2.6
Travel time (millions of passenger hours)	75,000	(+) 1,000	(+) 1.3

The 55 mph speed limit saves 2,000 to 4,000 lives per year, reduces highway fuel use slightly less than 2 percent, and saves taxpayers about $65 million per year (Table 1). But it also requires motorists to spend 1 billion additional hours driving each year, and the additional costs for enforcement are about $118 million per year. (Fines collected from speeders yield revenues that are roughly equal to these enforcement costs, so the net effect on government budgets is small.) ***Because of the substantial benefits to safety, the preponderant view of the committee is that the 55 mph speed limit should be retained on almost all of the nation's highways.***

Recommending some future course of policy for the 55 mph speed limit necessarily involves judgments about trade-offs between the various consequences as well as judgments about many other issues that surround this choice. This political choice, based on individual values, can best be made by elected officials. Such a choice may appear different for different parts of the highway system, however. For example, the 55 mph speed limit is relatively costly on rural Interstate routes, and in the discussion that follows the committee has highlighted the different balance between safety and travel time that applies to these routes.

Mechanism for Monitoring Compliance by States

The committee recommends that the federal government change the criteria it uses for monitoring speeds in each of the states to determine if each state is in compliance. The criteria currently in use do not adequately reflect safety priorities. To promote safety, federal law now permits the withholding of federal highway funds from a state in which more than 50 percent of all motorists exceed the speed limit on the roads posted at 55 mph. Adjustments for speedometer error and sampling error that are permitted under current law make this requirement less stringent than it first appears, but the committee finds the rationale and magnitude of these adjustments to be arbitrary.

The committee recommends that the adjustment currently allowed be discarded and recommends that the current method of measurement be replaced. The current method is fundamentally out of tune with conventional personal and organizational behavior for several reasons. First, courts rarely convict motorists cited for driving slightly above the speed limit. Second, police agencies traditionally have provided motorists some tolerance above the speed limit before initiating an enforcement action. Many police agencies have policies that allow officers to ignore small infractions. Only major infractions—10 mph above the speed limit in some jurisdictions—require the issuance of a citation. Third, motorists have come to expect less than literal interpretation of a posted speed limit. Despite the different degree of importance that individuals and organizations attach to speeding offenses, the current federal sanctions mechanism makes no explicit allowance for this American tradition. Nor does the current mechanism reflect relative safety priorities. The risk to safety is much greater for a driver traveling 80 mph than it is for a driver traveling 56 mph, but current federal procedures count both violations equally. ***To reflect both safety priorities and traditional organizational practices, the committee recommends that the federal government measure state compliance with the speed limit through a point system that attaches more significance to high-speed violations than to violations just above the speed limit.***

Such a point system could also be extended to reflect the different safety risks on various road systems. For example, it is more dangerous to drive 10 mph above the speed limit on a secondary road than on the Interstate system. Just as risks of very high speeds and moderately high speeds could be reflected in the number of points assigned to each vehicular speed monitored, so too could different points be associated with different road systems. In practice, this could involve relocating speed sensors or changing the procedure for computing average speeds so that it reflects safety priorities rather than energy priorities.

In addition, such a risk-related weighing scheme would encourage law enforcement agencies to allocate their resources so as to reduce the points counted against them. This means that proportionally more resources would be directed at the more hazardous offenders. Such a reallocation appears desirable in many instances and would redress one of the charges now made against the 55 mph speed limit, namely, that it encourages police agencies to misallocate resources on offenses that present relatively minor safety risks.

UNRESOLVED ISSUES

The committee was divided on the question of speed limits for rural Interstate routes. Approximately 31,500 miles of rural Interstate roads are posted at 55 mph; these routes represent about 6 percent of all mileage posted at 55 mph throughout the country, and they carry about 19 percent of all traffic that moves on roads posted at 55 mph. Disagreement among committee members stemmed from divergent views on several issues:

- Will growing noncompliance lead to the nullification of this law?
- Can the safety benefits of the 55 mph speed limit be sustained amidst the trend of increasing speed?
- Could safety be enhanced by permitting relaxation of the 55 mph speed limit if other effective safety policies are enacted to replace it?
- What would be the effect of permitting speeds somewhat higher than 55 mph on uncongested, high-quality roads?
- Could higher speed limits, if allowed on some roads, be contained to eligible roads?
- Does widespread noncompliance breed disrespect for law generally?
- Are the laws of each state a substantial deterrent to speeding?
- Can enhanced enforcement reverse the upward trend in speeding?
- Is control of speed limits a state or a federal responsibility?

These issues present choices that are best made by the Congress. Factual analysis of the historic and projected consequences of the 55 mph speed limit can only partly guide this decision. Elected officials will need to interpret the facts in the light of their perceptions about the broader range of issues involved. The committee's discussion of each of these issues, summarized in the following sections, may be useful to the Congress as it deliberates policy on the 55 mph speed limit.

Will Growing Noncompliance Lead to the Nullification of This Law?

As more states must rely on adjustments, permitted under current federal procedures to achieve technical compliance with the requirement that 50 percent of traffic complies with the speed limit, the likelihood increases that some states will be unable to demonstrate compliance, thereby subjecting themselves to a loss of some federal-aid highway funds.

Decreasing compliance, along with slipping public support and an apparent lack of support in some state legislatures, could ultimately lead to the nullification of this law. Advocates of the 55 mph speed limit argue that these developments are not harbingers of nullification, and that the technical adjustments allowed by current procedures are simply a means of adjusting the law realistically to operative enforcement norms. Others argue that the law is nearing the breaking point and that action should be taken now to preserve its most effective parts before the law as a whole has been irreparably weakened by public criticism and widespread noncompliance. Whether motorist unrest threatens the integrity of the existing law—a law that continues to be supported in concept by 76 percent of the public—is a matter on which individual views differ. Yet the question of whether the status quo can be sustained lies at the heart of current controversy about whether to repeal or modify the national 55 mph speed limit.

Can the Safety Benefits of the 55 mph Speed Limit Be Sustained Amid the Trend of Increasing Speed?

As average speeds increase, will the safety benefits remain as substantial as those noted earlier? A key to answering this question lies in whether the resulting traffic pattern has greater dispersion of speeds. When different vehicles travel at different speeds, driving involves more passing and lane changing, and safety is impaired. Statistical analyses often indicate that the variance in speeds, more than the average speed, is linked to safety. But variance declined when the average speed declined in 1974, and it is unclear whether variance can be held down as speeds increase.

Since 1974 the average highway speed has been increasing. The sketchy information that is available on the variance in speeds suggests that it may have remained down. But considering conditions before the enactment of the 55 mph speed limit, these recent trends indicate that about 60 percent of the reduction in average speeds that was achieved in 1974 has been retained. There is little statistical basis for estimating the effect on safety of these increases in speed, particularly as correspondingly large increases in variance do not appear to have occurred. In the absence of more reliable information, it is probably fair to assume that the erosion in safety benefits is somewhere between retrenchment in the average speed and the unaltered variance. Accordingly, the 55 mph speed limit may be about 20 percent less effective today than it was a decade ago.

Could Safety be Enhanced by Permitting Relaxation of the 55 mph Speed Limit If Other Effective Safety Policies Are Enacted To Replace It?

The committee recognized the immense practical difficulties of enacting a widespread behavioral change on the scale of the 55 mph speed limit, and it discussed whether the public determination could be mustered to introduce other laws of this significance. Recognizing the significant accomplishment of this law, there was strong sentiment in the committee that any relaxation of this law should be considered only if other equally effective policies could be substituted for it. Roads could be made safer and faster simultaneously. This might be achieved, in concept, by permitting states to increase speed limits above 55 mph on selected, clearly defined roads, but only if they took other actions that resulted in a net improvement in highway safety. Because of the belief that highway safety must continue to improve, the committee considered a variety of approaches by which relaxation of the speed limit could be linked to other safety policies. For example, a state might be permitted to increase the speed limit on its rural Interstate routes if it enacted a mandatory safety-belt use law. Nevertheless, no such recommendation resulted because the committee could not agree on a practical way to impose this concept. That is, it could not identify planning requirements or performance evaluations by which the federal government could be assured that highway safety would be improved in a state that raised some speed limits under this approach. Nonetheless, the concept of encouraging state action through incentives rather than punitive measures is appealing if practical ways can be found to achieve it.

What Would be the Effect of Permitting Speeds Somewhat Higher Than 55 mph on Uncongested, High-Quality Roads?

To illustrate the effects of allowing somewhat higher speeds (60 to 65 mph) on selected, uncongested, high-quality roads, the alternative of allowing increased speed limits on rural Interstate routes is discussed in this report. These routes are among the nation's safest roads, and the 55 mph speed limit exacts a higher price on them through additional travel time per life saved. Not all rural Interstates would be suitable candidates for higher speeds, and some non-Interstate freeways or controlled-access highways have safety and travel-time characteristics similar to those of rural Interstates. Nevertheless, for simplicity this discussion refers to rural Interstates.

Because Interstate routes have a better safety record than other road systems, relaxation of the speed limit on rural Interstate routes would probably not have a proportional impact on safety. If the speed limit were increased to 60 or 65 mph on all 31,500 miles of rural Interstate routes, about 19 percent of all traffic on roads posted at 55 mph would be affected. Projections indicate that this might result in about 500 more fatalities each year, an increase in consumption of about 10 million barrels of fuel annually, and an increase in taxpayer costs for programs affected by death, disability, or injury of about $10 million.

On the other hand, about 850,000 hours of travel time would be saved for every life lost. This means that about 100 years of travel time would be saved in exchange for the loss of one additional life, plus an additional serious, severe, or critical injury. This means that the cost of the 55 mph speed limit, measured in additional travel time, is about four times as great on rural Interstate routes, per life saved, as it is on the remainder of the highway system. Thus, relaxing the 55 mph speed limit on these routes would have a proportionally greater effect on travel time than it would on safety. One of the principal objections to the 55 mph speed limit is that applying the same maximum speed limit to all highways, without regard for their geometric design and safety performance, reflects an uneven commitment to safety across road systems.

The preceding estimates are based on an assumption that motorists on the different road systems behave independently. That is, the increase in speed limits is assumed to affect speeds only on those routes that are designated as eligible. This assumption may understate the full effect of increasing the 55 mph speed limit, as discussed next.

Could Higher Speeds, If Allowed on Some Roads, Be Contained to Eligible Roads?

If speed limits higher than 55 mph were permitted on some roads, this could alter driver behavior on other roads. When the 55 mph speed limit first became effective in 1974, safety improved on almost all road systems, even systems posted at speeds below 55 mph. The 55 mph speed limit may have changed drivers' perceptions of what speed was safe on all roads. Perhaps as motorists came to associate 55 mph as a safe speed on Interstate and primary highways, they unconsciously slowed down on other, less well-designed roads as well.

If spillover effects did occur in 1974, any modifications of the speed limit at this stage might likewise have spillover consequences. That is, authorization for speed increases on a small part of the system might produce system-wide changes in driving behavior. The strength of the spillover effect cannot be firmly established but it does provide one more reason for caution in contemplating upward adjustments in the 55 mph speed limit.

Similarly, if by increasing the speed limit on a few roads the federal government creates a public sense that the 55 mph speed limit on other roads is unimportant, slated for replacement, or unjustified, then the consequences of selectively increasing speed limits on a restricted group of roads could be greater than those detailed here. The predominant view of the committee is that such broader consequences can be avoided if, as part of any relaxation of maximum speed limits, an enhanced commitment is made to enforcement of the 55 mph speed limit on roads where it is retained.

Does Widespread Noncompliance Breed Disrespect for Law Generally?

Those who believe that the 55 mph speed limit should be relaxed argue that preoccupation with this particular statute impedes the enforcement of more vital laws, that it confuses the public about what is important in highway safety, and that it encourages misallocation of enforcement resources. Some argue that it is hypocritical to maintain the 55 mph speed limit while tolerating such widespread noncompliance, and that this could diminish public respect for other laws.

Advocates for retention of the 55 mph speed limit dispute whether this statute is different from other statutes with respect to compliance and the zeal with which it is enforced. Many traffic laws, such as coming to a full stop at stop signs, are widely violated. Even without perfect compliance, however, these laws significantly influence motorist behavior in desired ways.

Are the Laws of Each State a Substantial Deterrent to Speeding?

Part of the impetus for this study of the 55 mph speed limit came from congressional concerns about states that had reduced the penalty for violations of the 55 mph speed limit. Whether such practices have an effect on speeds and safety remains uncertain. Based on the level of fines authorized by state law and on whether points are assessed for speeding and entered into the offending driver's record, states were classified as high-penalty, medium-penalty, and low-penalty states. The low-penalty states as a group had higher speeds and higher fatality rates than the medium- and high-penalty states. But some of the individual states within the low-penalty group had lower speeds and lower fatality rates than many of the states in the other two categories. Further, there was no significant correlation between the severity of statutory penalties and the percentage of motorists speeding on Interstate routes per state.

Many other factors—actual police practices, court behavior, public awareness, and the like—influence how laws affect driving behavior. Both state laws and driver behavior may reflect regional attitudes. For example, in a state where motorists drive faster than the national average, the rate at which speeding citations are issued might increase and create public resentment. This in turn could lead to a legislative relaxation of penalties. In such an instance, the low penalties were imposed because of the amount of speeding that occurred; not the other way around. Because more speeding citations are issued per vehicle mile of travel in many of the low-penalty states than in other states, the low-penalty states may be facing particularly difficult compliance problems rather than shirking their responsibility for enforcement.

The committee concludes that there is no identifiable relationship between state statutes and compliance with the 55 mph speed limit. Rather, this linkage is immersed in a complex, interconnected set of relationships among laws, enforcement policies, court practices, and public attitudes.

Can Enhanced Enforcement Reverse the Upward Trend in Speeding?

Of all the actions that can be taken by enforcement officials, by courts, and through legislation to induce compliance with the speed limit, the high visibility of enforcement appears most effective. As in the case of violations involving drunken driving, severe penalties do not appear to be effective; they may generate counteracting reductions in citations and convictions. Although enhanced enforcement at specific sites has been demonstrated to increase compliance, the system-wide implications of this are uncertain, owing to the costs and practical constraints involved. Because nationwide there is only one on-duty state highway patrol officer, on average, for each 190 miles of road posted at 55 mph, there are clearly practical limits to how ubiquitous enforcement can be. Nevertheless, experimentation with different patroling strategies demonstrates that visible enforcement does make a difference in compliance although it may also impose additional costs. The implication for federal policy is that federal attention is more appropriately focused on the saliency of enforcement rather than on the severity of state laws.

Is Control of Speed Limits a State or a Federal Responsibility?

Aside from temporary federal action during World War II, speed limit policy had always been a responsibility of state and local governments before 1974. This changed because of national concern about energy shortages and the influence of vehicle speeds on fuel consumption. As fuel shortages have abated and safety has become the predominant issue in debates about the 55 mph speed limit, the federal role that has evolved is being called into question. Much of the debate about the appropriate role of state and federal government turns on philosophical differences that cannot be reconciled by analysis.

Nevertheless, the committee was impressed that the 55 mph speed limit has been one of the most effective highway safety policies ever adopted, and that its adoption was the product of a unique confluence of historic circumstances. The question today is whether to reject this legacy. The committee majority was reluctant to recommend doing so. It believed that, regardless of whether modifications of current law are desirable, it is appropriate for the federal government to continue to be involved in determining national maximum speed limits.

THE CHOICE AHEAD

Although the committee believes that the 55 mph speed limit should be retained on most highways and that changes should be made in the criteria by which the federal government monitors compliance, members were divided concerning how speeds should be set on rural Interstate routes. Some members favored retention of the 55 mph speed limit without exception; others favored some exceptions, generally on some rural Interstate routes; and others believed that any changes should be made as part of a broader safety improvement strategy that encourages states to enact other measures such as mandatory safety-belt use. The committee does not recommend any of of these specific steps. This is a choice that the Congress can best make, bearing in mind not only the safety, energy, and time consequences enumerated here but also the perceived necessity of change, judgments about the values of the various impacts, attitudes about appropriate state and federal roles, judgments about whether any relaxation of current law can be contained to the eligible routes designated, regional differences in public needs and attitudes, and concern about growing public noncompliance. The committee strongly believes that, in making this choice, the Congress should be aware of the substantial consequences of the 55 mph speed limit, the unique circumstances under which this change in policy was achieved, and the risks associated with any major modifications.

1 Introduction

For the past decade—since March 1974—a maximum speed limit of 55 mph has been in effect on the nation's highways. Originally enacted as a temporary fuel conservation measure, Congress made the 55 mph speed limit permanent because of the apparent safety benefits. In the year following the enactment of the 55 mph speed limit, the number of highway fatalities declined by about 17 percent, or by 9,100 fewer fatal accident victims than in 1973. However, a decline in travel as well as other factors also contributed to this improved highway safety record.

The 55 mph national maximum speed limit imposes certain burdens on society, especially in terms of added travel time and higher shipping costs. Because of these and other perceived burdens, several states have passed legislation urging repeal of the national maximum speed limit, or have greatly reduced the penalties for violations.

The issues at stake concern all Americans. Each year about 45,000 lives are lost in motor vehicle accidents and more than 3 million individuals are injured. Highway accidents are the leading cause of death among the nation's youth. More than one-half of all fatalities for those aged 16 to 19 are driving-related. The National Safety Council places the cost of the carnage on the nation's highways at more than $40 billion annually (*1*). Making highway travel safer is a top national transportation priority.

During the development of the Surface Transportation Assistance Act of 1982, proponents of the 55 mph speed limit proposed sanctions against those states whose penalties for speeding did not embody a substantial deterrent to violators. A compromise reached in the House and Senate conference requested that the National Research Council investigate the benefits of the law and determine whether the states with reduced penalties experienced more speeding.[1]

In response to this legislative mandate, a careful assessment is made in this study of the impact of the 55 mph speed limit on lives, injuries, travel time, fuel conservation, and taxpayer costs. Individual state laws and enforcement policies are examined to determine whether they affect motorist compliance with the 55 mph speed limit. The benefits of the 55 mph speed limit are analyzed alongside other policies and programs that have contributed to the nation's improved highway safety record since 1974.

Enactment of the national speed limit in 1974 represented a departure from historic public policy toward the regulation of highway travel speeds. Until 1974 the states had always in peacetime controlled motor vehicle speeds ever since New York and Connecticut passed the first laws imposing speed limits in 1901. Earlier, legal speed limits had adapted to how fast most motorists wished to drive. They had most typically been based on the speed attained by the 85th percentile of vehicles on a given roadway. As highways improved and as vehicle speeds increased, the states raised speed limits to reflect the speed the majority of drivers thought to be safe and reasonable.

Part of the current debate about the national maximum speed limit concerns whether the federal government should be involved in a matter that historically has been a state responsibility. Federal government involvement in controlling maximum permissible speeds is not entirely without precedent. In 1942 during World War II, the Office of Defense Transportation ordered a nationwide 35 mph speed limit to conserve rubber and gasoline. After World War II and until 1974, the federal government deferred to the states in setting speed limits.

Although the 55 mph speed limit has become predominantly a safety issue, it has far-reaching social and economic impacts on the lives of all Americans. Motorists drove more than 1.6 trillion vehicle miles in 1983. About one-third of their travel was slowed by the speed restriction. Slower speeds increase shipping costs, which ultimately affect the prices of consumer goods. These economic costs must be weighed alongside the potential safety and energy conservation benefits of the law as the Congress reassesses this issue.

SCOPE OF THE STUDY

The chapters that follow focus on how the 55 mph speed limit affects driving behavior and how that behavior, in turn, affects the national welfare. In Chapter 2 the changes in driving behavior brought about by the enactment of the 55 mph speed limit are described. Given the change in actual driving speeds on the various road systems, the effect of slower driving on highway safety is examined in Chapter 3. The effect of the 55 mph speed limit on safety in 1974, when the law was first enacted, is examined separately from its current contribution. Chapter 3 includes an examination of how improved emergency medical services, the shift to small cars, improved vehicle crashworthiness, increased truck traffic, demographic shifts, and other causes affect the long-term relationship between the 55 mph speed limit and highway fatalities.

An assessment is also made of how slower driving speeds reduced highway injuries, accidents, and the nation's medical bill (Chapter 4); lowered the cost of taxpayer supported programs affected by highway accidents (Chapter 5); and reduced motor vehicle energy consumption (Chapter 6). In addition to these positive impacts, the additional time drivers must spend making their trips is examined. In Chapter 7 particular attention is focused on geographical variations in the amount of additional driving time required and the impact of the speed limit on truckers and other frequent users of roads posted at 55 mph. Public opinion about the 55 mph speed limit is reviewed in Chapter 8. An examination of how recent changes in speed law enforcement and penalties have affected motorist compliance at the state level is included in Chapter 9. Following an estimation of the benefits and costs of the 55 mph speed limit, this report concludes with the recommendations of the committee and a discussion of the issues on which the committee could not agree (Chapter 10).

The committee examined considerable evidence to arrive at the conclusions and recommendations and much of this evidence is included in the appendixes to this report. The appendixes include data on recreational travel trends (Appendix A), a background paper on the effectiveness of mandatory safety belt use laws (Appendix B), a review of the speed and safety relationship based on recently available data (Appendix C), a summary of state laws relating to the 55 mph speed limit (Appendix D), estimating procedures used in some of the calculations (Appendixes E and G), and the results of a survey of state traffic enforcement administrators (Appendix F). In the course of this study, the committee identified a number of major gaps in knowledge about the relationship between speed limits and motorist behavior. Areas where further research is needed are noted in Appendix H. Detailed tables supporting technical analyses in the report are provided in Appendix I. Finally, biographical information on the committee is included at the end of the report.

The purpose of this report is to present the findings of the committee to aid the Congress in choosing whether to strengthen, weaken, or otherwise adjust the 55 mph speed limit. The alternatives to the current national speed limit include: raising the general speed limit, raising the speed limit selectively on certain roads or under certain road conditions, or changing the standards by which state compliance with enforcement of the national maximum speed limit are measured. Alternatively the Congress and the Administration could leave the 55 mph speed limit intact if, on balance, the need for change is not indicated by the evidence. These issues are discussed in both the Executive Summary and the conclusions chapter (Chapter 10).

REFERENCE

1. *Accident Facts*. National Safety Council, 1983 ed., Chicago, Ill., 1983.

NOTE

1. P.L. 97-424, Section 204, The Secretary of Transportation shall undertake to enter into appropriate arrangements with the National Academy of Sciences to conduct a comprehensive study and investigation of (1) the benefits, both human and economic, of lowered speeds due to the enactment of the 55 mile per hour National Maximum Speed Limit, with particular attention to savings to the taxpayers, and (2) whether the laws of each State constitute a substantial deterrent to violations of the maximum speed limit on public highways within such State. In entering into any arrangement with the National Academy of Sciences for conducting such study and investigation, the Secretary shall request the National Academy of Sciences to report to the Secretary and the Congress not later than twelve months after the date of enactment of this Act on the results of such study and investigation, together with its recomendations. The Secretary shall furnish to such Academy at its request any information which the Academy deems necessary for the purpose of conducting the investigation and study authorized by this section.

2
The Reductions in Speeds Resulting From the 55 MPH Speed Limit

As early as November 1973 a number of states began lowering posted speed limits as part of the national effort to reduce fuel consumption. By March 1974 all states had reduced maximum speed limits to 55 mph. Motorists responded positively to these efforts. Average speeds declined on all types of roads as Americans slowed down to save energy.

The effect of the new maximum speed limit varied considerably across the different highway systems. Major rural highways experienced the greatest reductions in posted speeds; many—particularly rural Interstates—had previously posted speed limits for automobiles of 70 mph or more. Two-lane rural highways and urban freeways experienced smaller reductions in speed because few had been posted above 60 mph before 1974.

The large-scale changes in driving behavior induced by the combination of the energy crisis and reduced speed limits have been sustained, for the most part, during the last decade. Some erosion has occurred in recent years as motorists have tended to drive faster. This may be explained by declining energy prices and fading memories of fuel shortages and long lines at gasoline stations. Nonetheless, average speeds on most highways remain well below those of the pre-energy crisis period. The effects of the speed limit on motorist behavior during the past 10 years are examined in this chapter, following an overview of the different types of highways affected by the 55 mph speed limit.

THE HIGHWAY SYSTEM: MILEAGE, TRAVEL, AND SAFETY

Each year Americans drive more than a trillion miles on nearly 4 million miles of streets and highways. This enormous amount of travel—enough to circle the globe 60 million times—is made up of a wide variety of trips. Individuals use their automobiles to travel to work, to shop, and for religious, civic, and recreational purposes. Businesses transport goods via truck and van across urban areas on local streets and across state lines on Interstate highways. The different roads and highways that make up the paved transportation network reflect the diversity of demands placed on them.

The Interstate system, characterized by multilane highways with traffic separated by direction and with controlled access, comprises only 1 percent of paved mileage but carries 20 percent of all traffic. Most long distance commercial and recreational travel occurs on the rural portions of this system. The urban segments also serve local traffic in and around major metropolitan areas. Almost all Interstates are posted at 55 mph today; most were posted at 65 to 75 mph in 1973. Despite the high speeds characteristic of this system, the Interstates are the safest highways; in 1982 they accounted for only 9 percent of motor vehicle fatalities (Figure 1).

Urban freeways, Interstate-type highways that are not part of the Interstate system, serve much the same function as urban Interstates. This is a small system, making up less than 1 percent of paved mileage and about 6 percent of total travel. About two-thirds of urban freeways are posted at 55 mph. Their safety record is comparable to that of the urban Interstates. Because urban freeways make up a small share of highway mileage and travel they are combined with the Interstates in Figure 1.

Arterial highways serve as traffic "arteries" by carrying traffic to and from urbanized areas. Almost all rural arterials, usually two-lane highways with wide lanes and shoulders, were posted at or above 55 mph before the energy crisis. Extensions of arterial highways in urban areas have typically had lower speed limits. Only 13 percent of urban arterials are posted at 55 mph today. Because arterials serve a key interconnecting purpose, they carry a large share of total travel. They constitute only 9 percent of all highway miles, but 43 percent of all travel occurs on these routes. Most motor vehicle fatalities occur on these roads as well (48 percent in 1982).

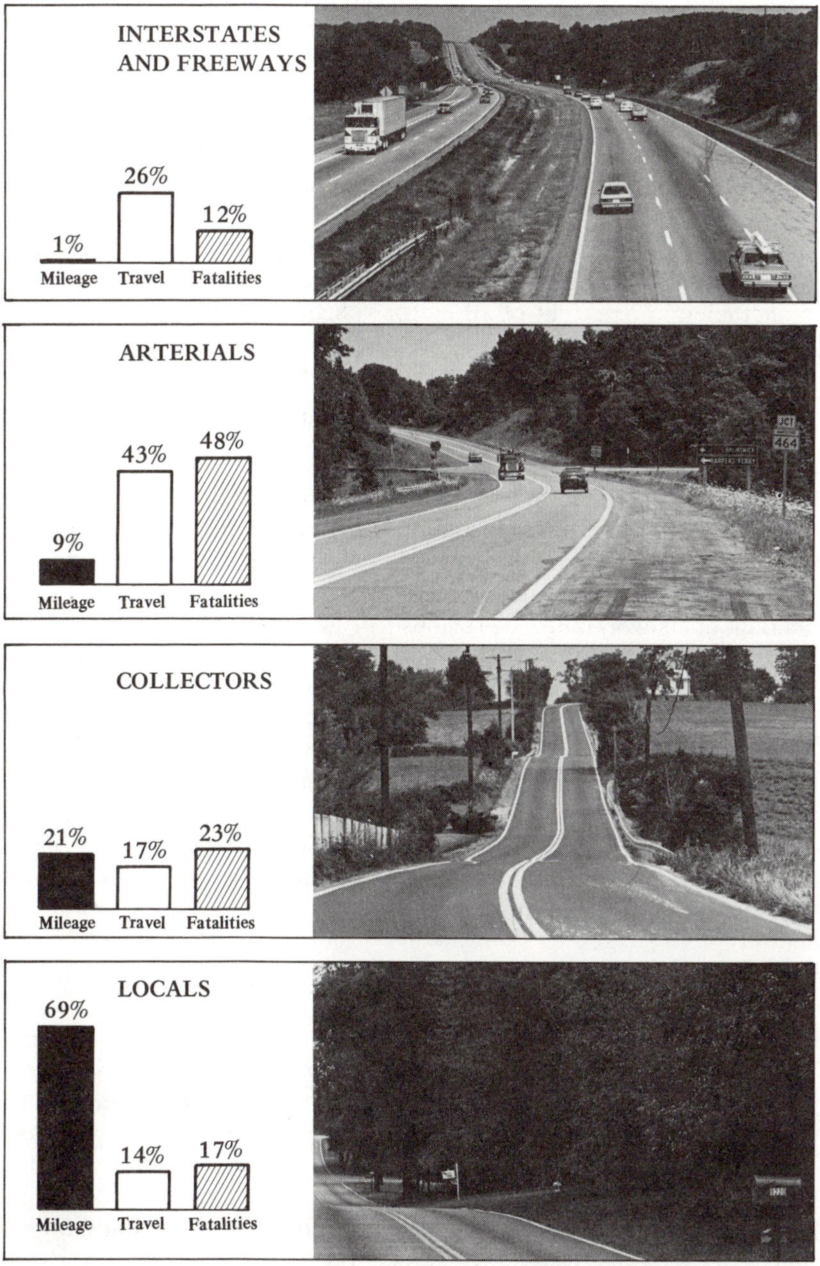

FIGURE 1 Roads and highways: mileage, travel, and safety (*1–3*).

Collector highways collect rural and urban traffic and funnel it toward arterial highways. They make up 21 percent of total mileage and carry 18 percent of all travel. Many rural collectors (38 percent) are now posted at 55 mph, and despite their lower design quality, some collectors were posted above 55 in 1973. These roads are the least safe highways posted at 55 mph, most likely because they have many street-level (at grade) intersections, limited sight distance, and other design limitations.

Local roads, neighborhood and residential streets, account for the majority of mileage (69 percent) but carry only 14 percent of all travel. Most of this is personal automobile travel at speeds of 15 to 30 mph. Nonetheless, 17 percent of all motor vehicle fatalities occurs on these routes.

AMOUNT OF TRAVEL AFFECTED BY THE 55 MPH SPEED LIMIT

When Congress made the 55 mph speed limit the national maximum, states lowered the posted speed limits on much of the Interstate system, on other freeways and expressways, and on many two-lane highways. Because few states had maintained records of posted speed limits, the amount of highway mileage previously posted above 55 mph is unknown. Based on estimates provided by the Federal Highway Administration, more than 577,000 miles of roads, about 15 percent of all highways, are posted at 55 mph, and these routes account for about 47 percent of all vehicle miles of travel (Figures 2 and 3). The speed limits on the majority of rural Interstates, arterials, and urban Interstates were probably reduced by the speed restriction. Many rural collectors and urban expressways were also affected. Some of these roads were always posted at 55 mph; thus it is likely that 35 to 40 percent of all traffic came under the new legal speed limit established in 1974.

SPEED CHANGES ACROSS HIGHWAY SYSTEMS

Except during World War II motorists had driven faster on main rural highways year after year until the early 1970s. As early as 1918 passenger cars were capable of traveling 30 to 35 mph on the open road, and by the beginning of World War II, average speeds on rural highways had probably increased to 45 to 55 mph. Wartime restrictions slowed traffic, but by 1947 motorists were again driving faster than ever, exceeding their prewar peak speeds and driving steadily faster by about 0.5 mph per year. By 1973 the average vehicle on rural Interstates traveled at 65 mph; more than 30 percent of vehicles traveled faster than 70 mph.

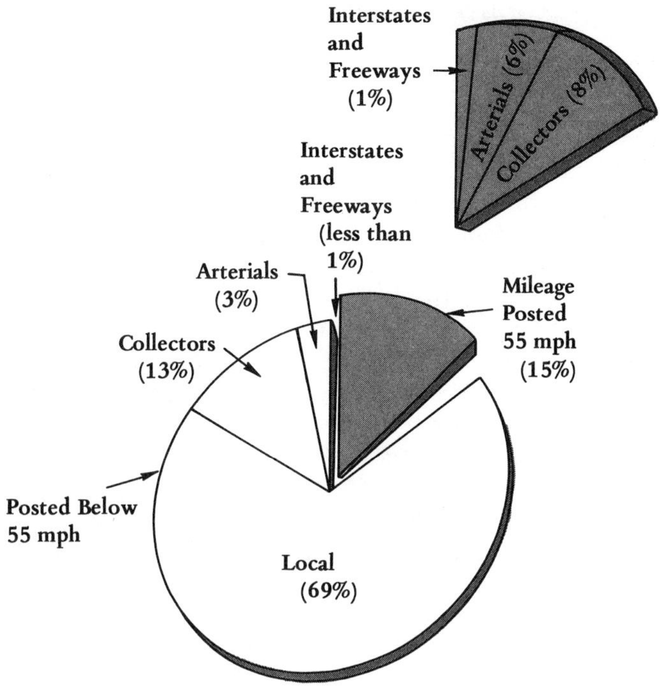

FIGURE 2 Highway mileage posted at 55 mph (1982) as a percentage of total mileage.

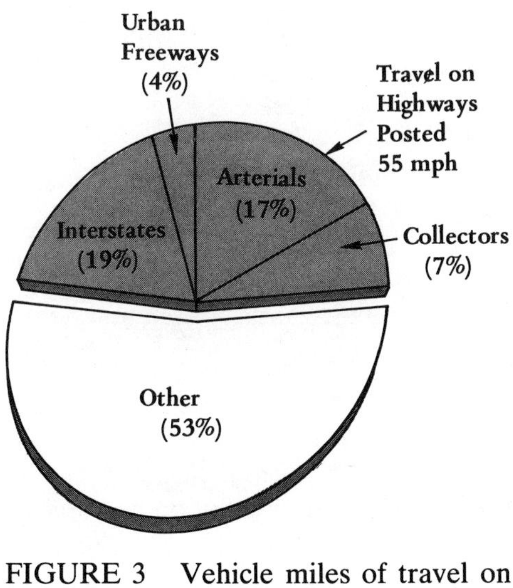

FIGURE 3 Vehicle miles of travel on highways posted above and below 55 mph.

Interstate Highways

The 55 mph speed limit caused a decrease in the average highway speeds on rural Interstates by dramatically curtailing the percentage of motorists driving at extremely high speeds. The percentage of drivers exceeding 65 mph on the rural Interstates declined from 50 to 9 percent during the first year the 55 mph speed limit was in effect. As a result the variance of speeds on Interstate highways narrowed considerably (see Figure 4).

In 1974 the average driver on the rural Interstate traveled at a speed slightly above 57 mph, down from 65 mph the previous year. Motorists have begun driving faster in recent years (59.1 mph in 1983), but these average speeds remain below those of 1973 (Figure 5). In addition, the sharp reduction in speed variance appears to have been sustained. Precise comparisons of the variance in speed over time are not possible because of changes in the devices used to measure speed and the manner in which data have been collected and recorded. Nevertheless, the best measure of speed variance (the estimated standard deviation) indicates a continued sharp discontinuity from pre-1974 levels (Figure 6).

A different pattern has developed on urban Interstates. As on the rural sections, motorists reduced their speeds in 1974, but since then

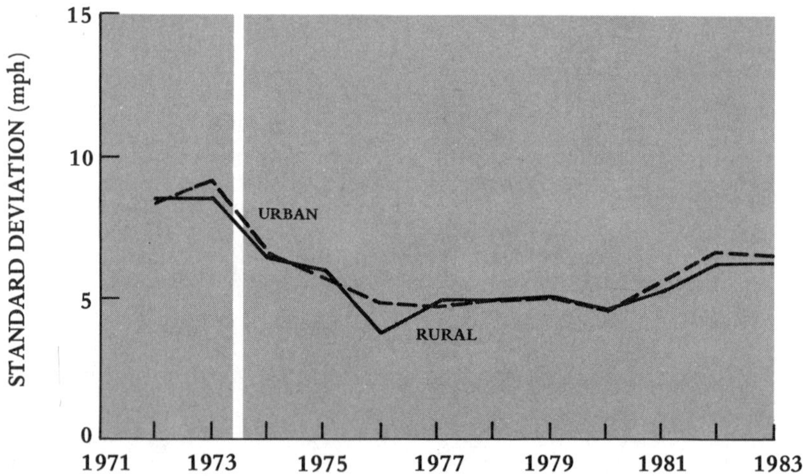

FIGURE 4 Change in speed distribution on rural Interstates, 1973–1974 (*4*).

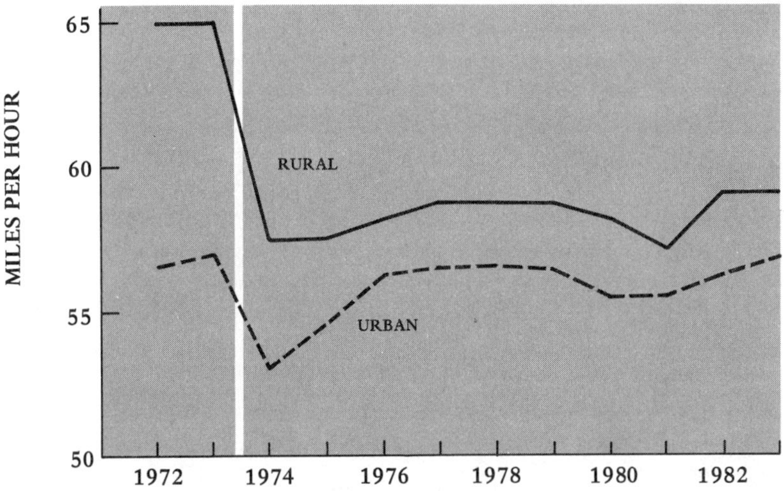

FIGURE 5 Trend in average speeds on Interstate highways (*5*).

average speeds have almost returned to the speeds of the pre-55 mph speed limit period. In contrast to the trend in average speed, the variance in speeds on these routes remains narrower than it was before the 55 mph speed limit was imposed (Figure 6).

Other Rural Highways

Motorists also slowed down on lesser rural highways. These highways were classified as primary and secondary routes in 1974 instead of the current designation of arterials and collectors. These classifications are not exactly equivalent. The federal-aid highway system was reclassified in 1976 and the definitions have shifted since that time.

The changes in travel speed between 1973 and 1974 reveal a shift in driver behavior similar to those on the Interstates. On both rural primary and secondary highways, average speeds declined because of the reduction in the percentage of drivers traveling at high speeds (Figure 7). This effect on rural secondary highways is even more surprising because speed limits were not reduced on all these roads. The 55 mph speed limit may have influenced driving speeds on highways posted below 55 mph. Drivers may have adjusted downward their perception of the appropriate safe speed for all roads when the highest quality freeways were limited to 55 mph (*6*).

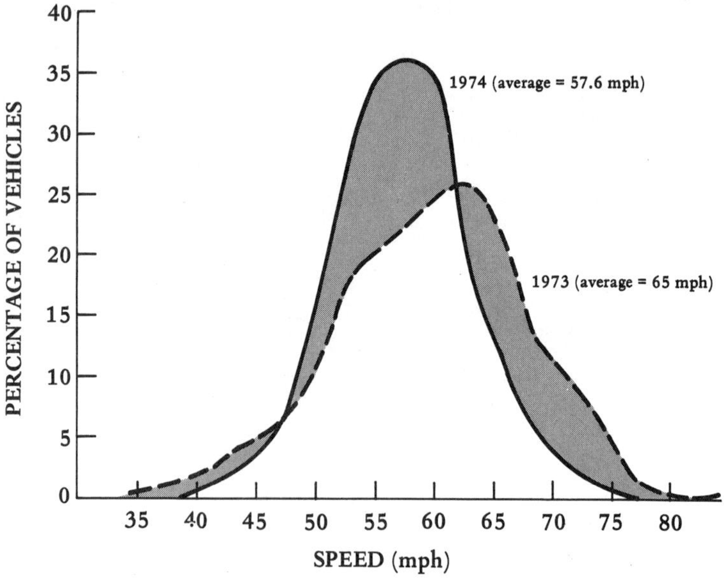

FIGURE 6 Trend in speed variability (standard deviation) on Interstate highways (*5*).

Current Speeds

Although speeds on most highways remain well below what they were in 1973, they have begun to increase. The increases in average speed are modest from year to year, usually less than 1 mph. Since 1981, however, statewide average speeds have increased each year. The changes between 1981 and 1982 may be overstated because of shifts in speed measurement techniques. Most states converted from radar for measuring speeds to automated devices imbedded in highway pavements. This interruption in the trend in speed measures is characteristic of the changes that have occurred to speed data over the years. These changes make precise statements about trends somewhat uncertain, as discussed in the following section.

PROBLEMS WITH SPEED DATA

Numerous changes have taken place in the ways in which speeds have been sampled, measured, and reported during the last decade. For example, in 1976, the federal-aid highway system was reclassified as a result of legislation. The shift in definitions of highways makes national trends inconsistent after 1976 for all highway classes except the Interstates. Fortunately speed data for the years surrounding implementation of the speed limit were consistently classified, thereby allowing observation of

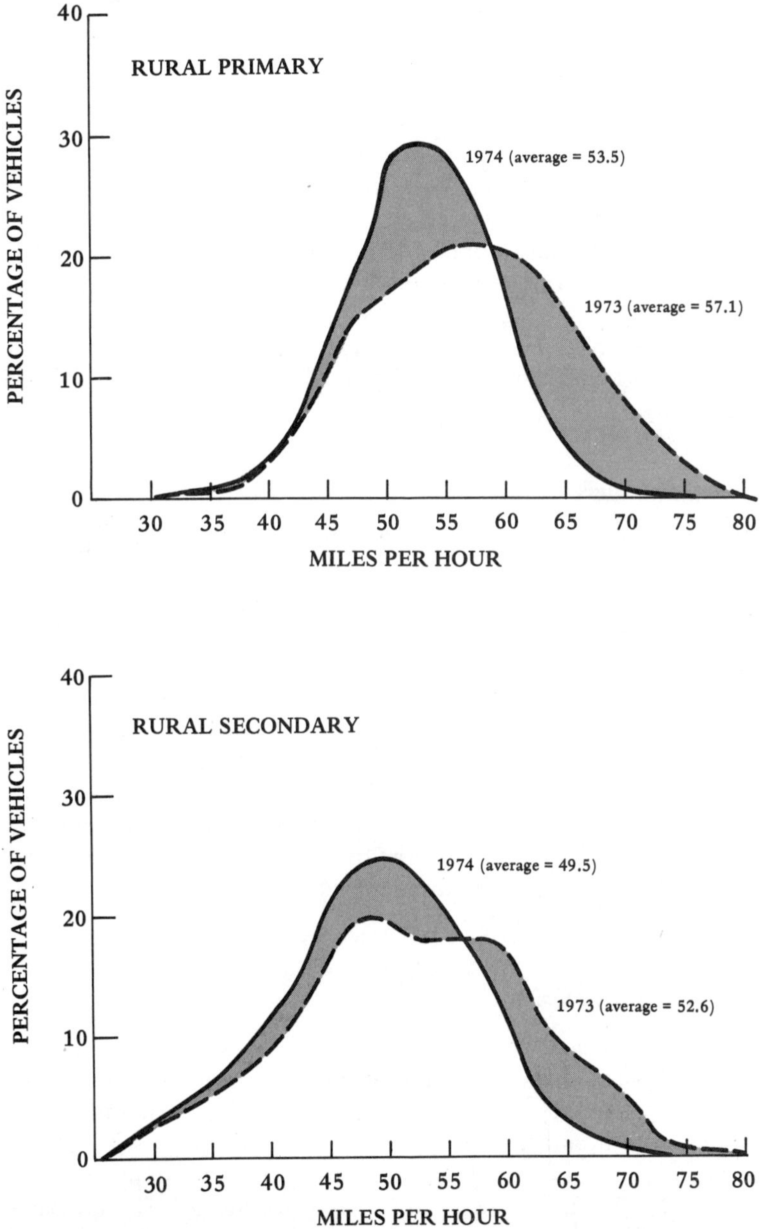

FIGURE 7 Change in speed distribution on non-Interstate rural highways, 1973–1974 (*4*).

the pronounced reductions in average speeds across all major highway systems.

Completeness of Sample

Before 1976 not all states reported speed data to the Federal Highway Administration (FHWA); thus these data are not strictly comparable on a year-to-year basis. Today all states are required to report speed data for a sample of roads representing each class of highway posted at 55 mph. Despite being an incomplete sample, the trend in average speeds in the pre-1976 period was remarkably stable from year-to-year (with the exception of 1974). This indicates that the trends were not disturbed by random fluctuations or by small changes in the number of states reporting.

Sampling Techniques

Until 1980 the states collected speed data without relying on a consistent sampling plan. Some states measured speeds on randomly selected highways and others did not. Speeds are now mostly monitored from fixed locations based on a prescribed sampling plan.

Before 1980 the states tended to measure the speeds of isolated vehicles under "free flow" conditions. This practice evolved from speed surveys designed to set speed limits based on the 85th percentile of free-flowing traffic unimpeded by congestion. Currently the speeds of all vehicles passing a sample site are monitored to provide an accurate measure of representative speeds on highways posted at 55 mph. Because most highways posted at 55 mph allow free flow conditions most of the time, this change in speed measurement techniques probably had only a marginal effect on speed trends.

Speed Measurement Devices

Automated devices measuring speeds with wire loops imbedded in highway pavements have replaced radar as the primary device for monitoring speeds. After testing these devices, the Federal Highway Administration found them quite accurate when properly used. Although some states began switching in the late 1970s, most states did not adopt the loop monitors until 1981. The increase in reported speeds in 1982 may have resulted from this shift in measurement techniques. Those states that switched to loop monitors in 1981 recorded a 6.4 percent increase in average speeds, whereas those states that already had loop monitors recorded an average increase in speeds of only 0.8 percent.

Loop monitors record higher average speeds than radar for two reasons. First, radar is difficult to conceal. Radar detection devices and CB radios warn motorists of the speed detection activities of the highway patrol. Motorists respond by slowing down. Loop monitors are more difficult to detect and less likely to be directly associated with speed enforcement than a patrolman holding a radar gun. Second, loop monitors also record night driving, and traffic often moves faster at night.

Reporting Periods

Until the mid-1970s speeds were usually recorded on 1 or 2 days during the summer to ensure that measurements were taken during good weather conditions. Summer is also the peak travel season. In 1976 the Federal Highway Administration began requiring quarterly speed reports to account for seasonal travel variations.

Because of all of the measurement changes, speed data are not perfectly consistent over time or within individual classifications. The extent of this inconsistency cannot be determined precisely, but even if each change has a small impact, the total effect might be large in some years. However, since the major changes in highway classification and sampling plans occurred in 1976 and in 1980, speed data trends before 1976, between 1976 and 1980, and for the years after 1980 can be viewed as reasonably consistent within these periods. The speed measures before 1980 do not precisely measure vehicle speeds, but, given the consistency in data gathering and measurement techniques, they portray a reasonably accurate picture of aggregate shifts in driver behavior within the periods noted previously. In particular, the sharp break with the trends in 1974 strongly suggest large-scale changes in driver behavior. The inconsistent trend data, however, limit their applicability in statistical models.

CONCLUSION

As a result of enactment of the 55 mph speed limit in 1974, average speeds on virtually all road systems were successfully reduced. The available evidence strongly suggests that the average declined because the highest-speed driving was sharply curtailed by the speed limit. Further, there may have been some spillover effects in that speeds declined even on some roads posted below 55 mph.

As public concern about fuel supplies waned, motorists began to drive faster. However, on rural Interstate highways average speeds remain well below pre-55 mph speed limit levels. The primary long-term effect of the 55 mph speed limit on other road systems has been the narrowed range of vehicle speeds. By reducing the amount of extreme high-speed driving, there is less passing and overtaking as traffic moves at a more uniform pace. As discussed in Chapter 3, the smaller variations in highway speeds is an important factor in reducing highway fatalities.

REFERENCES

1. *Highway Statistics, 1982*, Table HM-20. U.S. Department of Transportation.
2. *Highway Statistics, 1982*, Table VM-2. U.S. Department of Transportation.
3. *Fatal Accident Reporting System, 1982*, Table 17. U.S. Department of Transportation.
4. *Highway Statistics, Annual Issues 1972-1974*, Tables VS-1 and VS-2. U.S. Department of Transportation.
5. *Highway Statistics, Annual Issues 1972-1983*, Tables VS-1 and VS-2. U.S. Department of Transportation.
6. F.M. Council, L. Pitts, M. Sadorf, and O. Dart. An Examination of the Effects of the 55 mph Speed Limit on North Carolina Accidents. Highway Safety Research Center, University of North Carolina, Chapel Hill, 1975.

3
Highway Safety and the 55 MPH Speed Limit

During the Arab oil embargo of late 1973 and early 1974, the industrialized nations of the West experienced their first major shortage of petroleum. Many nations responded by implementing energy conservation measures such as reduced speed limits. Coincident with the introduction of the 55 mph speed limit in 1974, the fatality rate (deaths per 100 million miles of vehicle travel) in the United States declined 15 percent below the previous year. As a result of this decline 9,100 fewer individuals died in highway accidents. This remarkable reduction in the rate and number of fatalities is unparalleled in the peacetime history of the automobile age (Figure 8).

The U.S. fatality rate on all roads and highways has decreased substantially since World War II. The trend, though generally downward, fluctuates in response to major demographic and economic influences. Overall, however, highway safety has continually improved. Many factors contribute to these improvements. Motor vehicle standards have improved automobile crashworthiness (*2*). Today's vehicles include numerous safety devices unheard of 25 years ago. Modern highways are designed to separate traffic by direction, control access, and provide wider and more numerous lanes. The driving public has matured since the earliest days of driving when a license could be obtained with minimal driver training. Combined, these factors have led to a consistent decline in the fatality

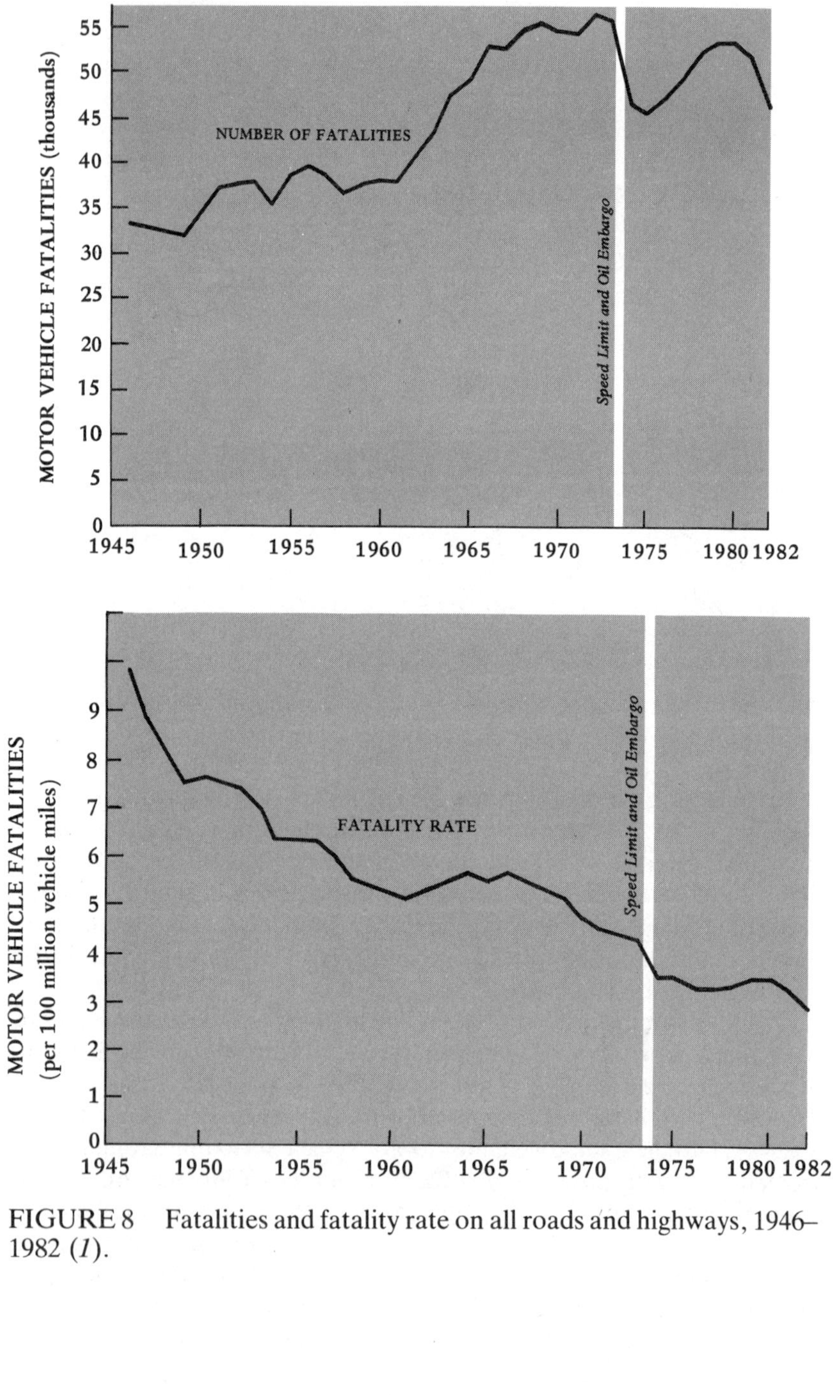

FIGURE 8 Fatalities and fatality rate on all roads and highways, 1946–1982 (*1*).

rate trend since 1946. Thus, even had the energy crisis not occurred, and had speed limits not been lowered in 1974, some decline in the fatality rate would have been expected in 1974 because of the prevailing downward trend. The fatality rate had declined from almost 10 fatalities per 100 million vehicle miles in 1946 to less than one-half that rate in 1973. A continuation of past trends would probably have yielded some reduction in the fatality rate between 1973 and 1974. The fatality rate had declined on average 3.1 percent annually between 1946 and 1973 (data provided in Table A-21, Appendix I). An average decline for 1974 would account for about 20 percent of the decline that actually occurred.

Other changes as well might have accounted for part of the exceptionally large decline in fatalities in 1974. The rising price of gasoline and the desire to save fuel reduced overall travel; 1974 was the first year since 1946 that total travel declined. The embargo may also have altered the types of trips that people made. Recreational travel, whether at night, on weekends, or for vacation trips, is generally less safe than normal travel for work and shopping. If such travel declined disproportionately, a disproportionate reduction in the fatality rate would be anticipated.

These alternative explanations are analyzed in detail in the first half of this chapter to isolate the contribution of the 55 mph speed limit to highway safety when it was introduced in 1974. The physical relationships between speed and crash severity and studies of speed and accident risk are first reviewed to provide a basis for understanding the speed-safety relationship. Many states, road systems, toll roads, and national statistics have been statistically analyzed to determine the effect of the 55 mph speed limit. These analyses and the supporting data and assumptions are reviewed and the estimates of the number of lives saved are compared. The experience of the United States with the reduced speed limit is also compared to other nations that experimented with lower speed limits during the energy crisis. Together, this experience indicates that the effect of the 55 mph speed limit on motorist behavior was a major contributor to the reductions in the fatality rate in 1974, accounting for 3,000 to 5,000 fewer fatalities that year.

In the second half of this chapter, the impact of the 55 mph speed limit on highway safety today, 10 years after it became effective, is assessed. The complexity of the many influences on highway safety, and the many years that the 55 mph speed limit has been in effect, make it difficult to estimate the current impact of the speed limit. In previous analyses various statistical techniques have been used to separate the effects of different influences on highway safety. These analyses are reviewed, as well as the potential offsetting effects of other safety policies, such as motor vehicle standards and drunk driving campaigns. The influence of economic and demographic factors on the number of highway fatalities is also discussed. Despite the partly offsetting effects of other safety improvements, the lower and more uniform speeds that have resulted from the 55 mph speed limit continue to save 2,000 to 4,000 lives each year.

SHORT-TERM IMPACT OF THE SPEED LIMIT ON SAFETY

Highway fatalities and the highway fatality rate declined in 1974, as shown in Figure 8, but these national trends include travel and fatalities on many roads that were not directly affected by the 55 mph speed limit. The speed limit applied to highways on which about one-third of travel occurs, and only about one-half of all motor-vehicle fatalities occur on highways posted at 55 mph. Unfortunately, the amount and speed of travel, as well as the number of fatalities, have not been consistently recorded for just those highway systems affected by the 55 mph speed limit.[1] The highway system with the most pronounced change in speeds, the Interstates, experienced the most pronounced change in fatality rate as shown in Figure 9. Whereas the fatality rate for all roads declined 15 percent, it declined 32 percent on the Interstates. In comparison, the fatality rate on local roads and streets, which are typically posted below 35 mph, showed no change (Figure 9).

Between these two extreme classes of roadway are major primary highways, secondary (usually two-lane rural) highways, and other urban and rural roads. Fatality rates declined less sharply on these systems than on the Interstates (Figure 10). Primary highways (not including the Interstates) and secondary roads had lower speeds than the Interstates and some were originally posted at 55 mph. The 17 and 13 percent respective fatality rate reductions on these highways in 1974 may be explained by this partial effect. Some road systems do not conform to this pattern, however. Federal-aid urban roads, for example, were probably posted below 55 mph before 1974; the average speed on this system was about 42 mph in 1973 (*4*). Although this highway system recorded a fatality rate reduction equal to those of the primary and secondary routes, the trend on this system is difficult to interpret. The highway mileage eligible under the then new federal-aid urban designation doubled between 1973 and 1974. Thus some of the improved trend may result from the inclusion of safer highways. Even so, pedestrian deaths on all roads and highways declined by 17 percent—a sharp reduction apparently not closely linked to the reduced speed limit. Nevertheless, most

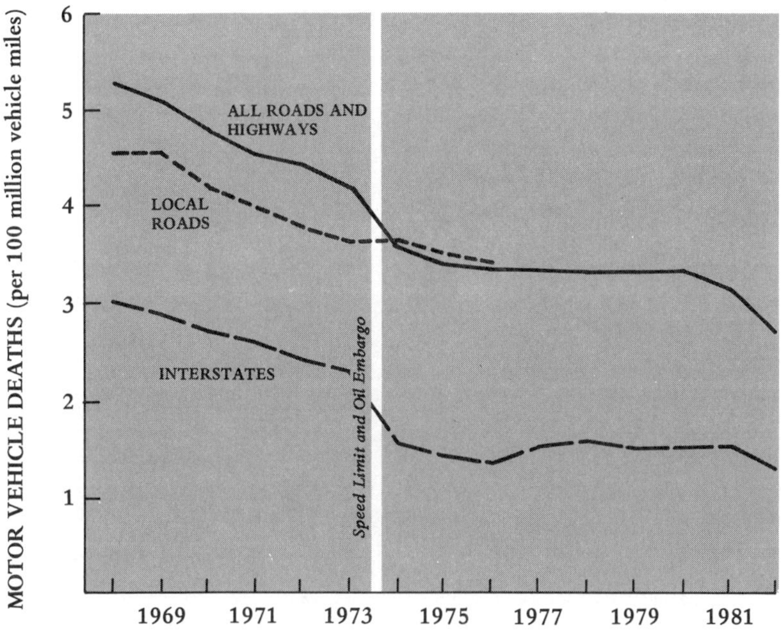

FIGURE 9 Fatality rates on all roads, Interstates, and local roads (*3*).

road systems in 1974 exhibited a decline in fatalities that corresponds roughly to the magnitude and extent of their speed-limit reductions.

Physical Reasons for the Effect of Speed on Highway Safety

Reduced speeds are likely to yield safer driving simply because of the physics of vehicles in motion; in particular,

- When traveling at a higher speed, a car moves a greater distance during the fixed period of time that it takes for the driver to react to a perceived problem;
- On highways lacking adequate superelevation, a driver's ability to steer safely around curves diminishes with speed;
- The distance required to stop a vehicle by braking increases with speed; and
- Crash severity increases disproportionately with speed at impact.

Each of these physical factors inherent in higher speeds affects highway safety, as discussed in the following sections.

Reaction Time

Modern highways are designed such that under optimal conditions a driver has at least 2.5 seconds after he perceives an accident situation well down the highway to decide to respond and to take corrective action (*5*). Vehicle speed does not affect reaction time, but at higher speeds a vehicle travels farther during a fixed perception-reaction time. For example, in 2.5 seconds, a car traveling 70 mph will travel 55 ft farther than one traveling 55 mph.

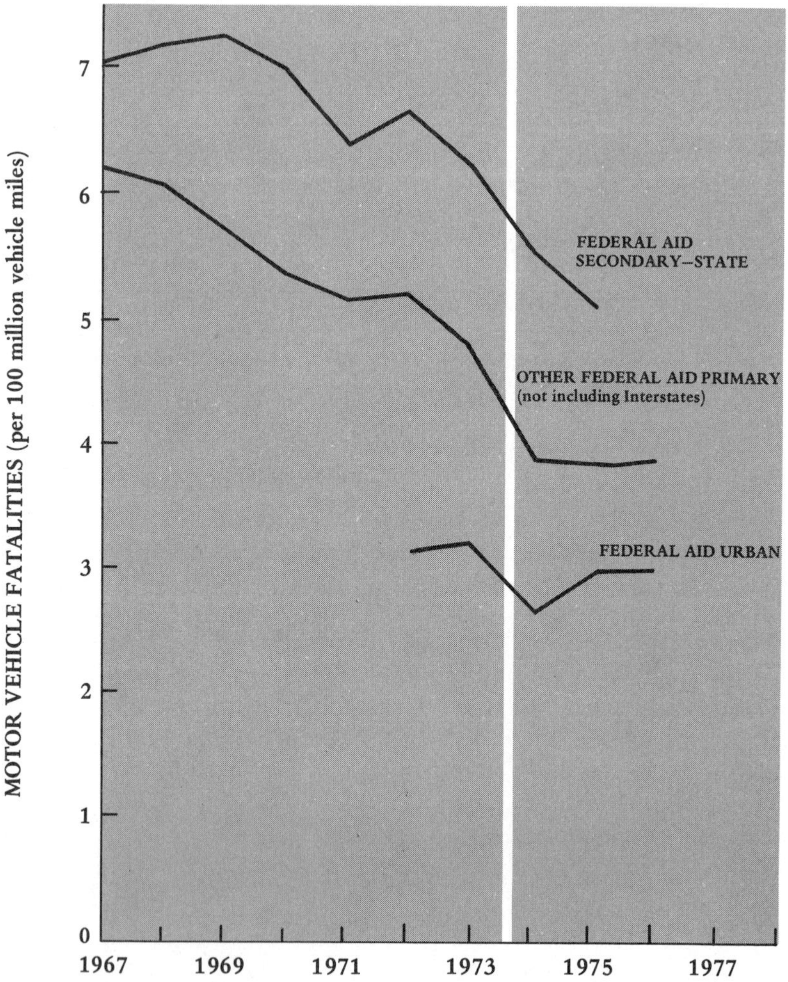

FIGURE 10 Fatality rate trends on federal-aid secondary, urban, and other primary highways (*3*).

Curves

Speed influences steering when negotiating curves on lower-quality highways. If a driver enters a curve lacking adequate superelevation at too high a speed, he may cut the curve, and thereby cross into the opposite lane. If he tries to stay in his lane, he may go off onto the shoulder.

Braking

The distance required to stop a vehicle increases disproportionately with speed. At normal braking deceleration (about 15 ft per second per second), a vehicle traveling at 70 mph travels 351 ft while the brakes are applied. A vehicle traveling at 55 mph continues in motion for 217 ft.

Impact Speed

The potential for injury severity increases at higher impact speeds at roughly a square power rate (*6*). In other words, the probability that an accident will result in an injury to the occupants increases faster than increases in vehicle speed at the time of impact. The speed of the vehicle at impact determines the amount of velocity change that a vehicle must undergo during a crash. Assuming that a vehicle strikes a fixed, unmovable object, such as a bridge abutment, it must decelerate in a fraction of a second. As the vehicle begins crushing into the abutment, the occupants continue their forward momentum into the windshield, dashboard, or steering wheel (assuming they are not wearing safety belts). Greater vehicle decelerations at impact disproportionately increase the probability that the occupants will be injured.

The importance of impact speed can be illustrated by referring to the formula that describes deceleration. A simplified version of this formula is

$$a = \frac{V^2}{64(s)}$$

where a is deceleration, V is the change in velocity from the time of impact until the vehicle is at rest, and s is the stopping distance in feet. In this formula the numerator (the change in velocity) increases as the square of impact speed. A reduction in impact speed of 5 mph (from 30 mph to 25—a 17 percent reduction) reduces the V^2 term by 30 percent.

These physical factors—reaction time, braking distance, and impact speed—interrelate in a nonlinear fashion. For example, comparing drivers traveling at 65, 60, and 55 mph, the fastest driver, on approaching a truck overturned ahead, will travel much farther while reacting, needs much more distance to come to a stop, and will crash into the truck at a higher impact speed.[2] Assuming a truck is blocking the road 290 ft ahead of these three drivers, the driver traveling at 65 mph will crash into the truck at 35 mph, would certainly be injured with such a great change in velocity, and could be killed. The driver traveling at 60 mph will crash into the truck at 22 mph and has a greater probability of surviving. The driver traveling at 55 mph would have sufficient stopping distance to avoid hitting the truck.

Whether motor vehicle occupants will be injured in a crash involves many more factors than stopping distance and impact speed. The type of object struck, the angle of the collision, the ability of the vehicle to absorb the energy being transferred during the crash, and a host of other factors influence the probability of injury severity in high-speed crashes. Indeed, tests conducted on human volunteers tightly strapped into deceleration sleds and on cadavers used in crashes have demonstrated that the human body can absorb remarkable decelerations if adequately restrained (*7*).

Speed contributes to injury severity, but the linkage is often complicated by numerous other contributing factors. For example, analysis of the causes of a single vehicle fatal accident on a two-lane rural highway might reveal that the driver had been drinking, that the vehicle had worn-out tires, that the driver entered a curve at an excessively high speed, and that the shoulder of the pavement had been eroded. With less to drink the driver might have slowed down on approaching the curve. Perhaps at a slower speed he might have managed the curve without hitting a rut. If the shoulder had been adequate, perhaps the car would have returned to the lane and avoided crashing. Had the vehicle struck an energy absorbing crash barrier rather than a bridge abutment the deceleration might have been gentle enough for the driver to survive. No single factor can be isolated as the principal cause of death in such an instance. Although the driver, type of roadway, vehicle condition, weather, lighting, alcohol impairment, and numerous other factors may cause or contribute to fatalities, studies of actual crashes (discussed next) corroborate the relationships between speed and injury severity.

Observed Linkage Between Speed and Safety

The National Crash Severity Study (NCSS), an intensive investigation of approximately 10,000 crashes from 1977 to 1979, revealed that the possibility of fatality increases dramatically as the change in velocity during the collision increases (*8*). In this study the change in velocity (Delta V) was assumed to occur over a constant 100 milliseconds beginning with the initial impact of the crash. The relationship between the probability of fatality and Delta V is shown in Figure 11. A driver crashing with a change in velocity of 50 mph is twice as likely to be killed as one crashing with a change in velocity of 40 mph (*8*).

Vehicle speeds also contribute to accident probability, particularly the variability in speeds on the same segment of highway. When motorists on the same segment of highway drive at widely divergent speeds, they run a greater risk of accident involvement (*9*). A wider variability in speeds increases the frequency of motorists passing one another which, in turn, increases the opportunities for multivehicle accidents to occur (*9, 10*). Examination of approximately 6,800 crashes on rural two-lane highways reveals that traveling 25 mph below or above the speed limit increases the probability of accident involvement approximately tenfold (Figure 12). The lower curve in Figure 12 shows a similar relationship based on more than 2,000 crashes on rural Interstate highways (*10*). Speed variability apparently contributes to the front-to-rear accidents prevalent on Interstate highways. These accidents are most common near intersections as motorists who slow down and change lanes to exit mix with motorists traveling at much higher speeds.

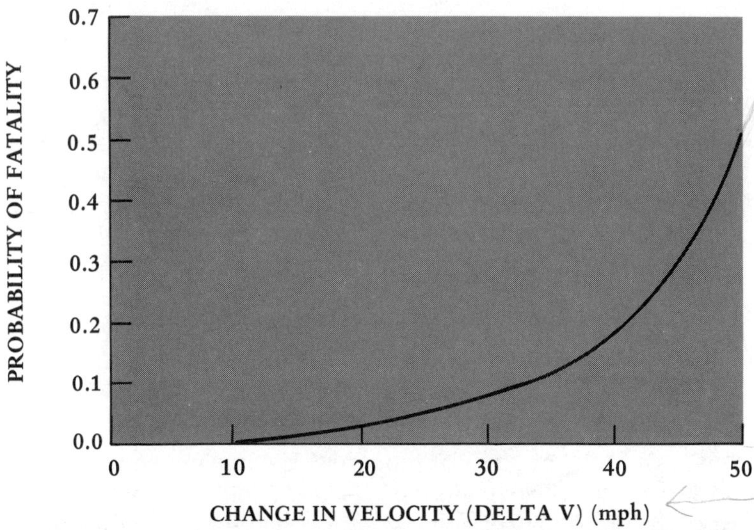

FIGURE 11 Probability of fatality for vehicle occupants related to change in velocity (Delta V) (*8*).

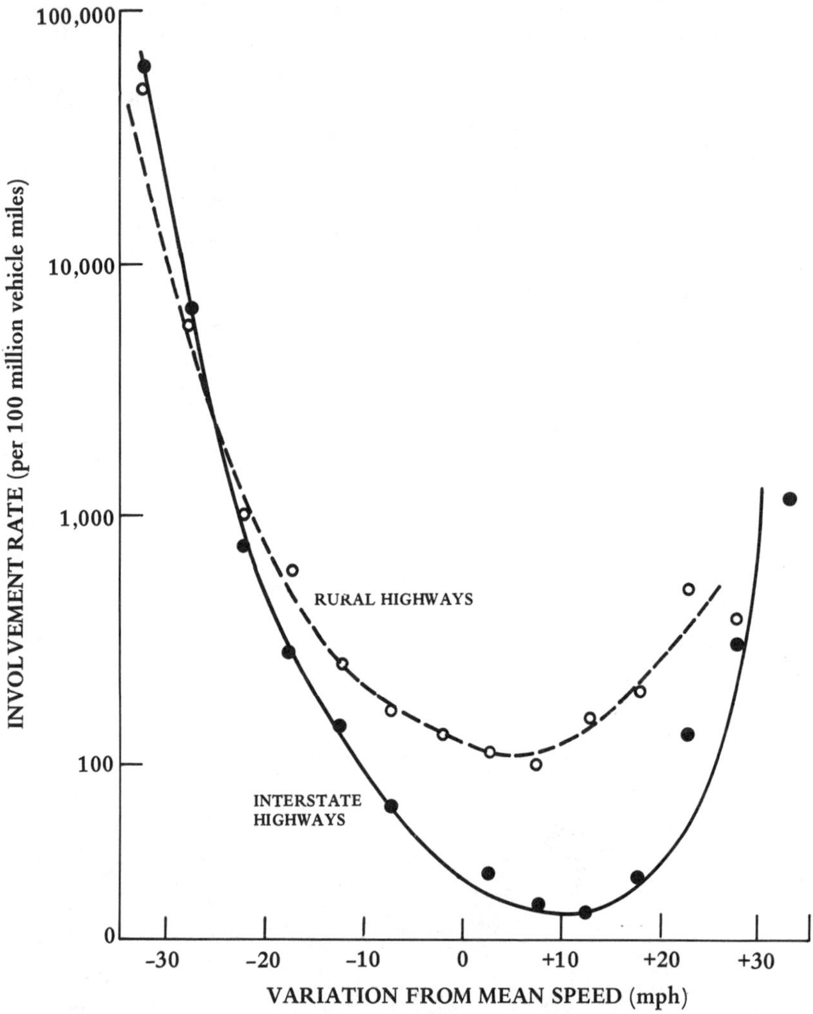

FIGURE 12 Accident involvement rate by variation from mean speed (*10*).

Although the relationship between the many causes of accidents and various speed-related physical properties is complex, empirical studies of crashes reveal a clear linkage (*8–10*). Wider variability in speeds increases the probability of accident occurrence (*9–11*). Crashes at higher impact speeds have a greater probability of resulting in a fatality (*6,8*).

NATIONWIDE EXPERIENCE WHEN THE 55 MPH SPEED LIMIT BECAME LAW IN 1974

Because speed is known to affect highway safety, based on assessment of the physical relationships involved as well as actual reported experience with accidents, it was expected that enactment of the 55 mph national maximum speed limit in 1974 would cause a decline in highway fatalities and injuries. Nevertheless, the magnitude of the decline—9,100 fewer fatalities in 1974 than in 1973—was so dramatic that renewed attention was given to the safety consequences of highway speeds.

Had pre-1974 trends continued, the fatality rate (deaths per 100 million vehicle miles) would have declined from 4.24 in 1973 to about 4.11 in 1974. In other words, if the average reduction in the fatality rate had occurred, the fatality rate of 1974 would have declined roughly 3 percent, far less than the 15 percent decline actually observed. Indeed, the decline in the fatality rate of 15.33 percent in 1974 is the largest decline experienced since World War II (Table A-21, Appendix I).

Simultaneous with the enactment of the 55 mph speed limit, travel was disrupted by the shortage of fuel and travel behavior was altered by the high price of gasoline. In a contemporary report the first 6 months of 1974 were compared with the same months of 1973 and the degree of difference attributable to various causes was also examined (*12*). On the basis of estimates of the magnitude of various causes, approximately one-half of the fatality decline, or about 4,500 lives saved per year, are attributed in this report to the slower and more uniform speeds on the highway.

The fatality rate observed on the Interstates and other major highways was significantly below past trends in 1974 and 1975 (*13*,*14*). One careful analysis of the potential impact of the speed limit revealed that it was probably the primary factor accounting for the reduction in the fatality rate, but no quantitative estimate was made because of data limitations (*13*). Another evaluation, which correlated the historic trend, reduced travel, and the speed limit, revealed that one-third of the variation in the fatality rate correlated with the historic trend and reduced travel. Two-thirds of the reduction in fatalities was attributed to the speed limit, which suggested that some 6,000 lives saved annually could be explained by slower and more uniform speeds (*14*). This estimate probably overstated the safety consequences, however, because it allocated the entire residual to the 55 mph speed limit.

In another comparison of trends from 1967 to 1972, using linear regression models, the individual contributions to safety of the historic trend, the speed limit, reduced travel, and a variety of other factors were identified (*15*). Fatal accidents were found to be significantly below historic trends on all rural highways in the primary and secondary systems (including Interstates, other major highways, and two-lane rural routes). On the basis of the sharp reductions in these trends and the significant relationships with speed changes, 35 percent of the fatality reduction was attributed to the speed limit in 1974, or about 3,200 of the lives saved.

More recently, monthly fatalities for the period 1970 to 1979 were analyzed by using a time series model to isolate the effect of the 55 mph speed limit (*16*). Monthly fatalities are not classified by type of road, however, making it impossible to focus exclusively on the highways with reduced speeds. This model of monthly fatalities was based on three independent variables: (a) a binary variable representing the speed limit, (b) an index of total exposure to risk (based on vehicle miles of travel), and (c) a measure of the improving trend in safety based on the declining fatality rate.

A model based on the first two variables (the speed limit and total travel) indicated a statistically significant relationship between fatalities and the speed limit policy, but estimated that the speed limit accounted for 10,400 fewer fatalities annually. The authors recognized that this estimate was too large; there were 9,100 fewer fatalities in 1974, and some of these occurred on highways unaffected by the 55 mph speed limit. To correct for this overestimate, the authors constructed a measure of the trend in safety based on the declining fatality rate. This resulted in a decline in the fatality reductions attributed to the 55 mph speed limit to roughly 7,500 annually. This estimate is based on the average fatality reductions for the post-1973 trend. The authors reduced this estimate still further because vehicle speeds and fatalities increased on highways posted at 55 mph in the late 1970s. The number of lives saved by the 55 mph speed limit was reduced to approximately 6,400 lives annually for 1978 and 1979.

Statistical attempts to unravel the events of 1974 based on national data have generally attributed to the enactment of the 55 mph speed limit an annual saving of lives between 3,200 and 7,500. Because of the shortcomings in the data, none of these estimates is definitive. The measures on the lower end of the range, however, appear more consistent with the data and the analyses reviewed elsewhere in this chapter.

A more precise numerical accounting of the effect of the speed limit on the highways most directly affected by reduced speeds requires longer fatality rate trend data for a longer period of time than are available for these systems. Because the data series begins in the late 1960s, the trends are too short to project the expected fatality rate for 1974 (independent of the 55 mph speed limit) with any great confidence. Some rough calculations, however, can be made with existing data. As noted earlier, the total fatality rate for the nation declined 15.33 percent in 1974. On the basis of the post-1946 trend, the overall fatality rate would probably have been reduced by about 3 percent that year independent of the 55 mph speed limit. The number of fatalities expected for 1974 (based on the amount of travel for that year and the projected fatality rate of 4.11 deaths per hundred million vehicle miles) suggests that of the 9,100 fewer fatalities, about 1,700 could be attributed to the improving fatality rate trend. This is about 19 percent of the reduction that occurred.

In addition, travel declined in 1974, but only by 1.45 percent. Applying the amount of travel in 1973 to the projected fatality rate for 1974 of 4.11 indicates that roughly 800 fewer fatalities (about 9 percent) occurred because of reduced travel.[3] Together with the reduced fatalities owing to the improving fatality rate trend, roughly 28 percent of the 9,100 fewer fatalities of 1974 can be accounted for by reduced travel and the trend in improved safety. On the highways most directly affected by reduced speed limits—the Interstates and rural primary and secondary highways—there were approximately 6,300 fewer fatalities beteen 1973 and 1974 (*17*). If this 28 percent reduction is applied to the 6,300 fewer fatalities on the highway systems with reduced speed limits, then about 4,500 reduced fatalities remain to be accounted for by reduced speeds and other factors. Although these adjustments are necessarily quite rough, they come closest to the studies of national data that suggest that the 55 mph speed limit saved from 3,200 to 4,500 lives in 1974.

OTHER CAUSES OF THE DECLINE IN FATALITIES IN 1974

Much of the difficulty in making definitive estimates of the safety consequences of the speed limit in 1974 arises from other events that occurred simultaneously. The Arab oil embargo, long lines at gasoline stations, weekend station closings, high gasoline prices, economic downturns, uncertainty of the availability of fuel, reduced travel, public attention to fuel conservation, media attention to the speed limit—these are but some of the changes that were occurring when the 55 mph speed limit became law. The behavioral changes of this period cannot be fully identified, much less measured and subjected to statistical analysis. Nevertheless, any attempt to isolate the effect of the 55 mph speed limit must operate in this confusing context.

Three travel-related aspects of this context are particularly important in assessing the relative contribution of the 55 mph speed limit:

- The extent to which highway travel declined in 1974,
- Whether accident-prone types of travel declined more than average, and
- Whether travel by accident-prone groups declined more than average.

Declines in Total Travel

Some decline in the absolute number of highway fatalities would have been expected in 1974 simply as a consequence of the reduction in travel. Total vehicle miles of travel (VMT) declined 1.45 percent nationwide in 1974, and travel declined on virtually every type of road. Thus, as noted previously, if the fatality rate per vehicle mile had remained unchanged, 800 fewer fatalities would have resulted simply as a consequence of the reduced traffic volumes. Estimates of total travel at the time, however, are derived largely from records of total fuel consumption and estimated vehicle fuel efficiency. These measures could have been distorted by the fuel crisis. If estimated travel was overstated, this would yield an understatement of the fatality rate, which contains the amount of travel in its denominator.

To verify VMT estimates, the average daily traffic from permanent automated counters, scattered throughout the nation on different highways, can be compared to estimated VMT. Automated counters provide a complete, continuous count of the vehicles traveling over them. They do not account for the distance traveled by each vehicle, and thus are not complete measures of travel, but they may provide an accurate index of changes in travel volume on the routes where they are installed.

Between 1973 and 1974 automated counters registered a 2.87 percent decline in travel. Some states experienced greater declines than others. Counts were down 4 to 8 percent during the first 4 months of 1974, but less thereafter. Traffic declined 4 to 5 percent on rural Interstates and non-federal highways. The data from automated counters do indeed suggest that fuel-tax-based estimates of VMT may be too high. Nevertheless, the discrepancy does not suggest any major error in the estimated fatality rate. For example, if VMT was really 1.42 percentage points lower than has been estimated (that is, assuming travel declined by 2.87 percent rather than 1.45 percent), then the decline in the fatality rate between 1973 and 1974 would have been 14.0 percent instead of 15.3 percent. This is still far beyond the range of previous trends and does not affect the conclusion that something unusual affected the pattern in 1974.

Furthermore, reliable toll road receipts for 1974 provide a further check on travel estimates because travel data from toll receipts are also derived from actual vehicle counts. The 22 toll roads with consistent data for the 1970s registered a greater decline in travel during both the 1973 and 1979 energy crises than recorded on the Interstates (*18*). Although Interstate travel declined slightly less than 2.5 percent, toll road travel declined 7.5 percent in 1974.

The toll road data appear to suggest that nationwide travel in 1974 declined more than measured by either automated counters or estimated VMT. If so, this would overstate the fatality rate decline in 1974 and reduce the estimate of the impact of the 55 mph speed limit on safety. This does not appear likely, however, because motor vehicle deaths declined on toll roads even more than on the Interstates. In 1974 the fatality rate on the Interstates declined by 32 percent whereas it declined by 38 percent on the toll roads. Speeds on toll roads and Interstates declined about the same (*19*). The fatality rate trends for these two highway systems are parallel between 1973 and 1979, as shown in Figure 13. This parallel trend indicates that estimates of travel on the Interstates are probably fairly accurate.

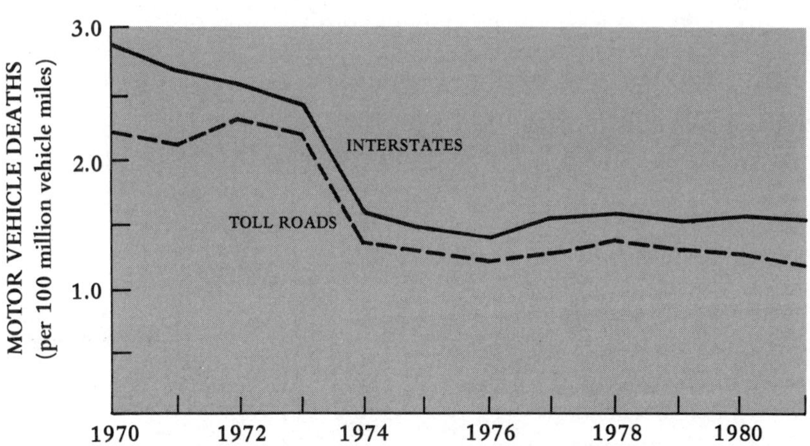

FIGURE 13 Fatality rate trends on Interstate highways and toll roads (*3, 8*).

Declines in Discretionary Travel

If individuals drove less on weekends, and at night, and took fewer vacation trips, fatalities would decrease because these types of trips involve above average fatal accident rates. National measures of such travel do not exist. Indirect measures and state and local estimates of discretionary travel, however, provide a rough picture of the effect of the energy crisis on these trips.

Inasmuch as discretionary travel occurs during some hours more than others, the available data do not suggest a major shift in the proportion of discretionary travel. For example, in 25 states with estimates of travel by time of day and day of week, travel declined in proportion to the decline in fatalities (*12*). Similarly, the percentage of fatalities occurring at night and on weekends actually increased in 1974 (Table 2), which suggests that the proportion of travel at these times was not reduced.

Vacation travel may have been curtailed more than average during 1974 and this could have affected the fatality rate. Vacation travel may be somewhat riskier than average travel because it often occurs on unfamiliar roads and highways, which may increase the probability that an accident will occur. Indicators of the amount of vacation travel in 1974 suggest that in some states vacation travel did decline. For example, in Washington, Florida, and New Mexico, measures of vacation travel declined more sharply than total VMT (Appendix A). This pattern does not appear to hold in all cases, however. For example, trips to national parks declined 1.6 percent in 1974, virtually the same decline observed in total travel (Table 3). Overall, the impact of reduced vacation travel on the fatality rate was probably relatively small. Long distance vacation travel during the early 1970s accounted for less than one-third of trips of more than 100 miles (*22*). In addition, the fuel shortages caused by the Arab oil embargo were concentrated in winter months, not in the months when most vacation travel occurs.

TABLE 2 Percentage of Motor Vehicle Fatalities on Friday and Saturday Nights, 1973–1976 (*20*)

Time	1973 (%)	1974 (%)	1975 (%)	1976 (%)
Friday				
10 pm to 1 am	17.1	21.2	22.0	19.5
Saturday				
10 pm to 1 am	17.8	20.0	19.8	19.0

TABLE 3 Visits to National Parks, 1971–1980 (000s) (*21*)

Year	Annual Total[a] (000s)	Percent Change
1971	192,527	–
1972	204,014	+5.6
1973	205,395	+0.7
1974	202,013	−1.6
1975	219,843	+8.8
1976	236,511	+7.6
1977	227,072	−3.9
1978	232,212	+2.3
1979	213,681	−8.0
1980	221,767	+3.8

[a]Parks not reporting consistently during the period were not included in the total counts.

Increased Travel by Accident-Prone Groups

An unusual decline in fatalities might also be observed if higher risk drivers, particularly males and young drivers, drove less. If they reduced their driving more than other groups, their representation in the fatality statistics would decline. In the case of young drivers exactly the opposite occurred (Table 4). The distribution of fatalities by age in 1974 as compared with 1973 indicates that vehicle occupants in the 16 to 24 year old age group actually account for a greater share of fatalities in 1974. These national statistics are not reported by age of driver; thus it is not possible to infer from these data that younger drivers drove less. However, analyses of more detailed state data reveal that deaths of teenage drivers increased as a share of total motor vehicle fatalities in 1974 (*13,23,24*).

TABLE 4 Age Distribution of Motor Vehicle Fatalities, 1973–1974 (percent of all fatalities) (*20*)

Year	Age					
	0–14 (%)	15–24 (%)	25–44 (%)	45–64 (%)	64–75 (%)	75+ (%)
1973	11.03	32.40	25.24	18.40	7.01	5.83
1974	10.51	34.28	25.50	17.50	6.62	5.31
Percentage change	−4.71	+5.51	+1.03	−4.89	−5.56	−8.92

Males, another higher than average risk group, also maintained a constant share of total fatalities from 1971 to 1976 (*20*). Further support for this lack of change is derived from a comparison of fatal accident victims in Arizona that revealed that driver characteristics—age and sex—remained constant between 1973 and 1974 (*25*).

Some of the improvements in safety in 1974, however, cannot be fully explained by the trend in safety, travel reductions, or slower speeds. Pedestrian deaths declined by 17 percent between 1973 and 1974, after having been relatively constant for the preceding 6 years. Because of less traffic on the road, and reduced speeds on most roads, some reduction in pedestrian deaths can be explained by these factors. Nevertheless, the full 17 percent decline remains unexplained.

Overall, the observed reduction in travel, possible declines in discretionary travel, and curtailment of travel by high-risk groups do not appear to explain much of the large decline in fatalities that occurred in 1974. Furthermore, well after travel resumed in 1975 and 1976, the fatality rate remained well below the levels of 1973. Had reductions in discretionary driving or other embargo-induced shifts in travel patterns caused the decline in the fatality rate in 1974, there would not have been a sustained reduction in the fatality rate after the fuel shortage abated and travel returned to its normal growth rate. In short, the various potential causes of the 1974 decline in fatalities, other than the 55 mph speed limit, simply do not explain the magnitude of the observed reduction.

Evidence of Speed Limit Impact at the State Level

Some of the ambiguities involved in isolating the consequences of the 55 mph speed limit can be reduced by examining state experience with this law. Data from some states are more detailed and more reliable than national data. For example, some states have records on the specific highways with reduced speed limits and others have detailed information about the reduced number of accidents that occurred in 1974.

The earliest studies of individual state experience simply compared the early months of 1974 with the early months of 1973 to determine whether safety improvements were due to reduced travel or reduced speeds. For example, in Arizona fatalities declined on all highways, but 92 percent of the decline occurred on high-speed roads, which implies that the speed limit was a principal factor in explaining the decline (*25*). In North Carolina and Michigan, however, fatalities declined as much on roads posted at or below 55 mph as on the Interstates (*23,24*).

In both these states average speeds and speed variance declined on all types of roads, not on just those posted above 55 mph. Two plausible hypotheses were offered: (a) motorists slowed down on all types of roads simply to save fuel, and (b) there was a spillover effect; that is, motorists coming off the Interstate after traveling only 55 mph slowed down proportionally on turning onto a two-lane highway, even though they were traveling below the posted speed.

Two separate studies addressed the improvements in safety in California by estimating the impact of such factors as the historic trend, reduced travel, reduced speed, and safety-belt use. The first attributed 46 percent of the decline in fatalities to the speed limit, and the later one, 39 percent (*26,27*). Assuming that California's experience was representative of the nation (California accounts for 10 percent of all travel) implies that the 55 mph speed limit saved between 3,500 and 4,200 lives.

Data from the state of Maryland permit the isolation of fatality statistics only for those highways where speed limits changed. A multiple-regression analysis of these data for 1970–1976 revealed that reduced travel accounted for about 8 to 10 percent of the fatality decline and the slower speeds accounted for 20 to 24 percent (*28*). If applicable at the national level, this means that 1,800 to 2,200 fewer fatalities per year can be attributed to the 55 mph speed limit. Despite a more sophisticated approach, these estimates remain rough because of the colinearity of the independent variables.

Studies of fatality declines in Texas and Illinois relied on time series models, which measured the impact of the 55 mph speed limit and the energy crisis with binary variables. Almost 60 percent of the fatality reductions in Texas and Illinois were attributed to the speed limit (*29,30*). Because of the binary variables employed by these models, their quantitative estimates of the impact of the speed limit remain approximations. The variable measuring the effect of the speed limit lacks sensitivity to speed changes. Nonetheless, the equations and the variables that describe the speed limit are statistically significant.

In Texas numerous safety indices were significantly related to the speed limit and the energy crisis (*31*). The fatality rate, fatal accident rate, injury rate, and injury accident rate declined well below historic trends when the 55 mph speed limit was introduced and remained below projections when the fuel shortage abated. The effects of the speed limit appeared to diminish in latter years.

Reductions in vehicle speeds played an important part in these safety gains. For example, fatal accidents in Texas in which the vehicle involved was severely crushed declined substantially between 1973 and 1974 (*32*). Vehicles involved in accidents are rated in some states on a seven-point scale, with a score of 7 being the most severe. Fatal accidents in Texas in which the vehicles involved were scored as a 6 or 7 declined 30 percent based on a comparison between the first 6 months of 1973 and the first 6 months of 1974 (*32*). Less severe crashes did not decline. Given the relationship between impact speed and crash severity, these statistics provide further support for the role of reduced speeds in explaining the safety improvements of 1974.

The speed, travel, and accident data from these state studies indicate that, generally, mean speeds declined more than 10 percent on Interstate highways. On non-Interstate roads, speeds declined 6 to 8 percent. The decline in travel varied across states; travel was generally lower on the Interstates than on the other systems. Fatalities, however, declined much more than travel on highways with reduced speed limits.

State experience with the 55 mph speed limit, as summarized in Table 5, indicates considerable consistency despite differences in regions and methodological approaches taken in estimating the effect of the speed limit on safety. All studies indicate a major improvement in safety as a result of reduced speeds. Care must be taken in evaluating these state results and in making nationwide inferences based on them. In some cases, extrapolating state results to the national level will yield an understatement of the likely impacts of the 55 mph speed limit on safety. For example, relatively small improvements in safety in North Carolina and Maryland could be traced to the speed limit, but these states had relatively modest reductions in posted speeds. In both states only the Interstate highways posted high speeds before 1974, and neither state has much Interstate mileage. On the other hand, extrapolation of state results could also result in an overstatement of the impact of the speed limit. For example, in Arizona most of the reductions in fatalities were attributed to the 55 mph speed limit, but the reduction in travel and the declining fatality rate trend were not sufficiently accounted for.

Among the other studies that made quantitative estimates of the impact of the speed limit on reduced fatalities, the findings range from 25 to 60 percent. This range, if applied to the fatality reductions at the national level, sugggest that the speed limit accounts for some 2,300 to 5,500 fewer fatalities.

INTERNATIONAL EXPERIENCE WITH SPEED LIMITS

The U.S. experience with the 55 mph speed limit, both nationwide and in those states where it was analyzed in detail, is consistent with similar experiences in other nations. Evidence from many nations during the past 30 years suggests that when speed limits are reduced, safety improves

TABLE 5 Estimated Effects of 55 mph Speed Limit in Individual States

State	Highway System	Time Period of Study	Methodology	Estimated Effect of 55 mph Speed Limit
Arizona (*25*)	Interstate, U.S. highways, state highways	1974 compared to 1973	Pre post comparison	34 percent decline in fatal accident rate attributed to 55 mph speed limit
California (*26*)	All highways with changed speed limit	First quarter of 1974 compared to preceding 5-year trend	Linear regression	46 percent of decline in fatalities due to reduced average speed and speed variance
California (*27*)	All highways with changed speed limit	1974 compared to preceding 10-year trends	Linear regression	39 percent of decline in fatalities
Illinois (*30*)	Affected highways compared to unaffected	Monthly data 1971–1977	Time series models	60 percent of reduction in fatal accidents (1974 compared to 1971–1973 average)
Indiana (*33*)	All highways with changed speed limit	1974 compared to preceding 3 years	Analysis of variance	Effect of speed limit on 67 percent reduction in fatality rate not estimated
Kentucky (*34*)	All highways	1974 compared to 1973	Pre post comparison	Speed limit key cause of reduction in accident rate
Maryland (*28*)	All highways with changed speed limit	1970–1976	Multiple regression	21 to 24 percent of fatality reductions explained by speed limit

TABLE 5 (continued)

State	Highway System	Time Period of Study	Methodology	Estimated Effect of 55 mph Speed Limit
Michigan (*35*)	Major state highways	Monthly data 1972–1974	Multiple regression	38 percent of reduced fatal accidents on major state highways and 8 percent on Interstate attributed to 55 mph speed limit
Michigan (*24*)	All highways	1974 compared to 1973	Pre post comparison	Effects of speed limit in reducing fatal accidents apparent in second half of 1974
North Carolina (*23*)	All roads and highways	1974 compared to 1973	Pre post comparison	Reductions in fatal accidents not explained by declines in travel—speed limit important in reducing speed variance but effect of 55 mph speed limit on accidents not estimated
Texas (*29*)	Affected highways compared to unaffected	1972–1974 monthly data	Time series models	57 percent of decline in fatalities in highways with reduced speed limits attributed to 55 mph speed limit
Texas (*31*)	All highways	1968–1975 monthly data	Time series models	Reductions in accident rates indicate safety improvements due to reduced speed limit
Utah (*36*)	Affected highways compared to unaffected	1971–1975 monthly data	Analysis of variance	Differences in fatal accidents statistically significant on affected highways, but not significant on unaffected highways

(*37*). Experiments with speed limits in Sweden and Finland during the last two decades indicate that reduced speed limits led to a decline in average speed and speed variance, which resulted in reduced numbers of injury accidents (Table 6).[4] By comparing the accident experience of comparable roadways where speed limits were unchanged, Swedish studies verified that the speed limit caused the safety improvement. Roadways with reduced speed limits consistently recorded larger safety gains than the comparison roads, which usually showed little change (*38, 39*).

With the advent of the Arab oil embargo a number of nations temporarily reduced speed limits to conserve fuel. New Zealand, Finland, Australia, Denmark, and the United Kingdom reduced posted speeds and examined in detail their experiences with accidents (Table 7). In all cases motorists responded to the urgings of their governments to drive slower, and the frequency of injury accidents and fatalities declined. Because some of these nations do not systematically record highway traffic volume, accident rates could not always be computed. However, indicators of travel in New Zealand did not suggest major reductions in travel, certainly not in proportion to the overall decline in fatalities (*43*). The dramatic reductions in fatalities in New Zealand and Australia cannot be explained by reduced travel alone; the sharply reduced fatalities indicate that the fatality rate declined as well.

As the energy crisis waned in 1974, motorists began driving faster in Great Britain and Australia (*45,43*). As a result, these nations subsequently lost the safety gains achieved through lower speed limits. In New Zealand, however, speeds remained below historic levels, as did the number of fatalities and injuries (*43*).

Changes in speed limits have also been examined in Ontario (*44*), France (*41*), South Africa (*41*), and West Germany (*42*). In line with the experiences of the other nations given in Tables 6 and 7, these countries also experienced an improvement in safety when speed limits were reduced.

Although international experience with speed limits is not always recorded in sufficient detail to permit development of quantitative relationships between reductions in posted speed limits, changes in vehicle speeds, and reductions in fatalities and injuries, it nevertheless suggests a significant relationship between reduced speed limits and overall safety. If the relationship derived from reductions in speed limits and improved safety abroad is applied to the United States, it implies that between 2,900 and 5,000 lives would be saved in the United States as a result of the 55 mph speed limit (*48*). Although this estimate is necessarily rough, it is consistent with the findings from state studies, which imply a saving of 2,300 to 5,500 lives if these findings are extrapolated to the national

TABLE 6 Experiments with Speed Limits Before Energy Crises of 1974 and 1979

Jurisdiction	Speed Limit (mph) pre	post	Highway Type	Years	Effect on Safety	Comment
Finland (*40*)	FS[a]	54[b]	All	1962	8 percent reduction in accidents	Temporary speed limit reductions in some areas of country; travel data not available
	FS	54	All	1966	16 percent reduction in accidents	
	FS	54	All	1968	13 percent reduction in accidents	
	see Comments—right margin			1970–76	Fatal accidents declined by 25 to 50 percent	Results of reductions in speed limits on various routes throughout country
France (*41*)	FS	60	Major routes	1969	40 percent reduction in accidents	
	FS	60	Major routes	1970	Rate of injury accidents declined 14 percent	
West Germany (*42*)	FS	60	Rural roads	1972	Fatalities declined by 9.8 percent, injury accidents by 9.6 percent	Autobahn injury accidents (unaffected by speed limit) declined by 8 percent in same period
Sweden (*38,39*)	FS	54	Two lane	1966–69	Rate of accidents declined 16 percent	Effects on comparison roads very small in comparison
	FS	60	Major routes	1966–69	Rate of accidents declined 4 percent	
	66	54	Two lane	1969–70	Rate of accidents declined 30 percent	
	54	42	Two lane	1970–72	Rate of accidents declined 22 percent	
	78	66	Two lane	1978	Rate of accidents declined 31 percent	

[a]FS = Unposted speed limit.
[b]Kilometer per hour limits converted to miles per hour.

TABLE 7 International Experience with Reduced Speed Limits in Response to Fuel Shortages

Jurisdiction	Speed Limit (mph) pre	Speed Limit (mph) post	Highway Type	Years	Effect on Safety	Comment
Finland (*40*)	72 to 66	48	Major routes	1973–1974	Fatalities declined 30 percent	
France (*41*)	FS	54	Major routes	1973	23 percent decline in accidents	Safety-belt use law passed simultaneously, but would not affect number of accidents
New Zealand (*43*)	55 to 60	50	Rural roads	1973–1974	Fatalities declined 37 percent Injuries declined 22 to 24 percent	Travel estimates did not show proportionate reductions
Ontario (*44*)	70	60	Freeways	1976–1976	Rate of Fatalities declined 22.8 percent	Safety belt law passed simultaneously. Fatality rate fell twice as much on high speed-roads.
	60	55	Other		Fatal accident rate declined 28.7 percent	
South Africa (*41*)	72	48	Rural roads	1973–1974	Accident rate declined by 48 percent	
Sweden (*38*)	66	54	Two lane	1978–1979	Fatal accident rate declined 52 percent, injury rate 25 percent	Comparison roads had only a 6 percent decline in fatality rate
United Kingdom (*45*)	70	54	Motorways	1973–1974	Injury accident rate declined 42.5 percent during the day and 27.5 percent at night	
	60	50	All purpose	1973–1974	Injury accident rate declined 16.5 percent during the day and 19.1 percent at night	
	70	50 to 60	All purpose	1974–1975	Injury accident rate declined by 11 percent	
Victoria, Australia (*46*)	70	60	Rural roads	1973–75	Fatal accidents declined by 24 percent, injury accidents declined by 14 percent	
Denmark (*47*)	FS[a] FS	48 to 66[b] 48 to 54	Motorways	1973–1974	Fatalities declined 20 percent	Effect of reduced travel factored out

[a]FS = Free speeds, unposted speed limit.
[b]Kilometer per hour limits converted to miles per hour.

level. It is also consistent with the studies of national data, which generally estimate that as a result of the 55 mph speed limit fatalities were reduced by 3,200 to 7,500 lives in 1974. Although no single source of data or analytical interpretation can yield an unequivocal number of lives saved as a result of the 55 mph speed limit in 1974, most analyses indicate estimates in the range of 3,000 to 5,000 lives saved.

EFFECT OF THE SPEED LIMIT AFTER ONE DECADE

Despite considerable agreement about the effect of the 55 mph speed limit in 1974, its influence on safety a decade later raises many additional questions and uncertainties. As with many other policies that attempt to modify public behavior, the benefits of the 55 mph speed limit may have diminished along with declining public awareness and compliance with the law.

As noted earlier, highway safety generally has been improving, year by year, for the last half century. Although the same physical relationships govern reaction time, steering control, braking distance, and vehicle deceleration during impact, the effect of speed may nevertheless have been reduced by other improvements to highway safety during the last decade. Modern vehicles can more successfully withstand high-speed crashes, improved roadside designs have made various features less dangerous, advances in emergency access to medical treatment can prevent unnecessary deaths for some accident victims. Because of such improvements, some of the initial effect of the 55 mph speed limit may have been offset. In other words, as further improvements are made to roads, vehicles, and medical services, motorists are offered the same protection from risks that is afforded by driving 55 mph. One aspect of such overlapping protection is that if speed limits were increased to pre-1974 levels, the effect on safety would be smaller than it would have been a decade ago.

In addition, there is less compliance with the 55 mph speed limit today than there was in 1974. To the extent that adherence to the law has slipped, so too might the effectiveness of the law have diminished.

To answer the question: "How effective is the 55 mph speed limit in saving lives now?" three related questions are explored in this section:

- How do current highway speeds compare with those of 1974?
- Have improvements in other roadway, vehicular, or medical features substituted for the original effect of the 55 mph speed limit?
- Do the trends in highway fatalities since 1974 indicate that the 55 mph speed limit has had a permanent effect?

Speeds in the 1973–1983 Decade

Although speeds have increased during the last decade, speeds on main rural highways remain well below the levels of 1973. Average speeds on rural Interstates declined by 7.4 mph in 1974. Even though they have increased by 1.5 mph since that time, 80 percent of the reduction in average speed has been retained (calculations given in Appendix I, Table A-22). Average speeds on other major rural highways declined by 3.6 mph in 1974. The speeds on these highways also have increased by 1.1 mph since 1974, yet 70 percent of the reduction in speed caused by the 55 mph speed limit still exists.

In contrast to rural speed trends, average speeds on the urban Interstates have returned to near pre-1974 levels. Currently speeds are only 5 percent below what they were in 1973. Although most of the speed reductions on these highways induced by the 55 mph speed limit have been eroded, urban Interstates and freeways account for less than one-third of all travel on highways posted at 55 mph. Almost 70 percent of all travel on highways posted at 55 mph occurs in rural areas.

In addition, speed variance on highways posted at 55 mph appears to have remained fairly constant between 1974 and 1982. Precise comparisons of the variance in speed for all highway systems are rendered impossible by reclassification of the federal-aid highway system in 1976 and by changes in the devices used to measure speed, and the sampling techniques from which estimates of speed are drawn. Nevertheless, the best available estimate of speed variance on Interstate highways (as measured by the standard deviation) indicates a sharp discontinuity from pre-1974 levels (Figure 14). Thus, overall, the speeds of most of the

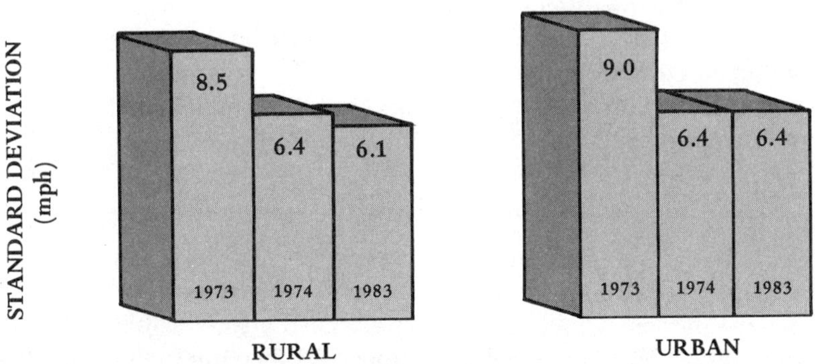

FIGURE 14 Estimated standard deviation in speeds on Interstate highways, 1973, 1974, and 1983 (*4*).

nation's travel on high-speed roads remains well below the levels that prevailed before the 55 mph speed limit.

Other Improvements to Roads, Vehicles, Medical Services, and Driver Behavior

Some of the safety benefits that accompanied the introduction of the 55 mph speed limit in 1974 might not be lost if speed limits were increased today, because improvements during the last decade make high-speed driving less risky than it formerly was. Improvements in vehicles and highways, and demographic and economic changes, have contributed to the trend in increased highway safety during the past decade. Virtually the entire automobile fleet currently in use has been manufactured to the vehicle standards implemented in the late 1960s. Additional improvements to the highway system are made every year. Demographic changes, particularly the shrinking number of teenage drivers, contribute to safer driving. Other safety polices, such as crackdowns on drunk driving, may also have important effects on safety.

Other influences, however, appear to be negative. An increasing percentage of the vehicle fleet is made up of smaller cars that offer less protection in a crash. Although speeds remain well below the levels before 1974, they have been creeping upward.

Of all the changes in the highway environment that occurred during the past decade, seven changes appear to have the most important influence on highway safety today; these changes are examined in more detail in the following sections.

- Improved emergency medical services.
- Crackdowns on alcohol-impaired driving.
- Physical improvements to highways.
- Introduction of more stringent motor vehicle standards.
- Increased use of smaller cars.
- Increased volume of heavy truck traffic.
- Fewer teenage drivers, more elderly drivers.

Improved Emergency Medical Services

The quality of emergency medical services, such as accident notification, ambulance services, and hospital care, has improved during the last decade. Emergency medical services are currently available to approximately 30 percent of all accident victims (*49*). Federal funding for these activities increased from $1.7 million in 1968 to $12.7 million in 1981, and states spend approximately $3 for every $1 spent by the federal government.

Although no comprehensive evaluation of the benefits of improved emergency medical services has been made, reports from individual jurisdictions note substantial improvements in safety due to improved ambulance services (*50,51*). A comprehensive emergency response and treatment system in West Germany may well be contributing to the reduced fatality rate on the Autobahn (*52*).

Because of these improvements, individuals involved in high-speed crashes may have a better chance of surviving today than in the past. Overall, however, these improvements cannot replace the improvement in safety that was generated by the 55 mph speed limit, because currently the improved emergency medical services are available to only about one-third of accident victims. In addition, many high-speed crashes occur in rural areas that have considerably less access to medical centers.

Crackdowns on Alcohol-Impaired Driving

Many governments, in the United States and abroad, have taken special steps to deter drinking drivers. Among these are more severe penalties, legal changes in the type of evidence of drinking admissible in court, and enforcement blitzes of various types. These efforts have significant effects, but they do not last (*53*). While public attention is focused on the problem, these efforts save many lives; individuals are less likely to drive under the influence of alcohol, which may mean that they can operate a vehicle more safely at higher speeds. Over the longer term, however, permanent social changes required by laws of this nature are difficult to sustain. When media and public attention wanes, the changes in behavior they induce dissappear (*53*).

Recent legislation in some states to increase the drinking age may also lead to improved safety. States that lowered the drinking age in the mid-1970s had an increase in the number of motor vehicle fatalities involving teenagers (*54*). Although this policy may improve safety, its effect will fall almost entirely on drivers aged 18 to 20. Thus, the overall effectiveness of the 55 mph speed limit is largely unrelated to this particular policy.

Physical Improvements to Highways

Modern highways, designed for high speeds, offer far safer driving than earlier roadways. The separation of intersections has reduced the frequency of right-angle crashes, which are frequently fatal for the occupant of the vehicle struck. By separating the direction of traffic with medians and barriers, fewer head-on crashes occur. In recent years the ability of roadside barriers to absorb impact force has been improved; obstacles in medians have been removed; and breakaway sign posts have been installed. Highways of Interstate design quality, which incorporate many of these improvements, have safety records two to four times better than other roads.

As highways become safer, they become more forgiving of driver mistakes. For this reason, the effectiveness of the 55 mph speed limit, relative to when it was first introduced, may not be as great. Although the size of this overlapping effect cannot be readily quantified, it probably contributes to the steadily declining fatality rate on high-speed highways, particularly Interstates and urban freeways. On two-lane rural highways posted at 55 mph, fewer of these improvements have been made. One-half of all travel on highways posted at 55 mph occurs on arterial and collector highways, which are generally built to standards well below the standards of the Interstates. Indeed, arterials and collectors account for 75 percent of the deaths on highways posted at 55 mph. On these particular highways, therefore, the benefits of the 55 mph speed limit have probably not diminished.

Introduction of More Stringent Motor Vehicle Standards

Beginning in the mid-1960s motor vehicle manufacturers were required to phase in vehicle safety improvements to reduce the risk of injury and fatality. The standards called for installation of safety belts, energy-absorbing steering columns, head restraints, and side-door beams among other improvements (*55*). Evaluations and review of the earliest requirements suggested that energy-absorbing steering columns and side-door beams reduced the risk of fatality in accidents (*2,56–58*). (The benefits of safety belts are estimated separately in Appendix B.) Subsequent evaluations of energy-absorbing steering columns and side-door beams further documented the reduced risk to fatality, and head restraints were shown to reduce the risk of injury (*55*).

A recent attempt by Graham and Garber (*59*) to model the impact of motor vehicle standards on fatality trends suggests that these improvements in aggregate save many thousands of lives each year. Although their estimates are admittedly crude, they do support the findings from evaluations and engineering studies, which estimate that more stringent motor vehicle standards reduce the risk of fatality by 15 to 35 percent.

These overall improvements in vehicle standards indicate that the occupants of today's mid-sized cars would have more protection in the event of a crash than in a similar car of the mid-1970s. As a result, the estimated benefits of the 55 mph speed limit, in terms of lives saved, is partly replaced by the benefits resulting from these standards. As discussed in the next section, however, the benefit depends on the size of the car. The shift to smaller and lighter cars erodes some of the value of improved crashworthiness.

Increased Use of Smaller Cars

The Arab oil embargo of 1973–1974 induced a shift in consumer preferences for smaller vehicles with greater fuel efficiency. In 1974 smaller cars made up about 30 percent of the total vehicle fleet (as defined in Figure 15). This share will increase to about 45 percent in 1984 (*60*). Although more economical to operate, smaller cars offer less protection in the event of a crash. The drivers of small cars—whether small car is defined in terms of weight (900 kg) or category (small subcompact)—are twice as likely to be killed than drivers of large cars in the event of a crash (*61*,*62*).

The overall safety risk of smaller cars has not yet been fully identified. Some statistics indicate that these vehicles are involved in an above average number of crashes (*63*), particularly rollover crashes, but other statistics suggest the opposite (*64*). When involved in an accident, however, small vehicles appear to be at a distinct safety disadvantage. For example, in crashes between smaller and larger cars, the occupants of smaller

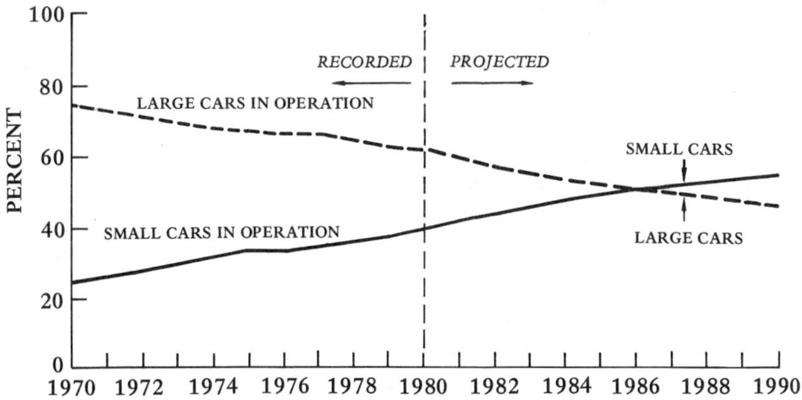

Large Cars = Mid-sized and large

Small Cars = Two-seater, minicompact, subcompact, and compact

Size categories are based on Environmental Protection Agencys interior roominess classification.

FIGURE 15 Small and large cars in operation, 1970–1990 (*60*).

cars have a much greater probability of being seriously injured or killed. Occupants of subcompact cars involved in crashes with larger vehicles are from 6 to 8 times more likely to be killed than occupants of other cars (*60*). Although the risk of small-car and large-car crashes will diminish after about 1986 as the vehicle fleet shifts to smaller vehicles, based on the projection shown in Figure 15, by 1990 smaller cars will represent only 55 percent of the vehicle fleet.

Given these disparities in vehicle mass, and the exponential contribution of impact speed to crash severity, the 55 mph speed limit plays an important role in reducing the severity of crashes for smaller cars. Although today's subcompact is more crashworthy than 1974 models, motorists would probably be safer in a standard-sized car of the mid-1970s. For these reasons, some of the gains of improved crashworthiness are nullified by the growing number of smaller cars.

Increasing Risk of Crashes Between Cars and Heavy Trucks

Although large trucks have a better overall accident record per mile traveled than cars, when accidents occur the outcomes are more likely to be fatal (*65*). Large trucks are involved in only 6 percent of crashes but account for 12 percent of fatal accidents (*66*). A principal contributing cause to these severe outcomes is the disparity between the mass of trucks and cars. A fully loaded truck (weighing 80,000 lb) outweighs a standard-sized car by 20 to 1. In the event of a truck-car crash, the car will most likely sustain a greater change in momentum. In 1982, for example, of 3,935 fatalities that resulted from multivehicle accidents involving trucks weighing more than 26,000 lb, 84 percent of the victims occupied the other vehicle (*67*).

The Federal Highway Administration projects that travel by combination trucks will increase 63 percent by 1990 whereas automobile travel is projected to increase 30 percent (*68*). Because of increased heavy truck travel and the shift to smaller cars, the frequency of small car/heavy truck crashes will probably increase (*66*). Slower impact speeds help reduce the severity of these type of crashes; thus the 55 mph speed limit has an important role to play in future highway safety.

Fewer Teenage Drivers, More Elderly Drivers

The number of young drivers on the road has declined 2 percent since 1979 whereas fatalities in this age group have declined by 8 percent. Like the crime rate, the highway fatality rate has decreased with the aging of the baby boom generation. Drivers 16 to 24 years old have about three times as many fatal accidents per mile traveled as those aged 25 to 64 (*69*).

Other shifts in the age distribution, however, offset the decline in the number of young drivers. For example, drivers above age 65 have increased 6 percent since 1979, and although they typically drive fewer miles, their estimated total accident involvement rate is about twice that of drivers in the 25 to 64 year old age group (*69*).

The approximate impact of these changes on the number of fatalities can be estimated by comparing the expected number of deaths for these age cohorts at two points in time. The only influence on the number of fatalities would be the change in the relative size of the age groups. Overall, the change in the age distribution reduces the number of fatalities by about 450 when comparing the age distributions for 1973 and 1982 (calculations given in Table A-23 of Appendix I). Although a clear benefit, this reduction is about 1 percent of total motor vehicle deaths, which suggests that the net effect of demographic shifts is relatively small from year to year.

As the population ages, the speed limit may increase in importance as a safety policy. Older persons are more susceptible to injury when crashes occur, and crashes at high speeds are more likely to be fatal for these drivers. With the elderly population growing the fastest of all age groups, the 55 mph speed limit will grow in importance as a safety policy.

Net Effect of Safety Improvements: 1974–1984

During the past decade, and despite the upward creep in average speeds, the fatality rate has decreased from 3.59 in 1974 to 2.76 in 1983. This 23 percent reduction reflects all of the specific behavioral, vehicular, or roadway changes examined previously. It also reflects other shifts in the composition of the motoring public, changes in public awareness of safety, or any other factors that may have altered the safety of the highway system since the 55 mph speed limit was first introduced.

Altogether, these additional improvements have probably induced changes that partly duplicate the protection afforded by the 55 mph speed limit. Increases in the speed limit today would probably have less of an impact on safety than they would have had a decade ago. Nonetheless the 55 mph speed limit still appears to have a substantial impact on safety. Simply adjusting the life saving effects of the 55 mph speed limit to account for the increased travel and improved safety since 1974—as well as accounting for the upward creep in average speeds—suggests that 2,000 to 4,000 fewer fatalities occurred in 1983 as a result of the 55 mph speed limit.[5]

Statistical analysis, discussed next, generally supports an estimate of this magnitude, although estimating what the nation's highway safety record would be without the speed limit necessarily involves speculation about numerous changes in the highway environment that have occurred since 1974, many of which were discussed earlier. Nevertheless, the current effect of the 55 mph speed limit on safety can be estimated statistically from trends in speeds, fatalities, and other key components of the highway environment.

Statistical Estimate of the Current Effect of the 55 mph Speed Limit

Statistical evidence indicates that the 55 mph speed limit remains an important safety policy, even though the effect of each of the myriad factors contributing to improvements in highway safety during the past decade cannot be estimated simultaneously with multivariate models. Many of the changes that have occurred cannot be or have not been measured. Further, some of them interact with the effects of speed, which makes it impossible to separate their effects statistically.

The policy of reduced speeds continues to indicate a statistically significant relationship with both fatalities and the fatality rate. A model of monthly fatalities for the period 1970 through 1979 (discussed in the first half of this chapter) indicates that the intervention variable for the speed limit policy was related to the trend in fatalities (*16*). Although this model has technical limitations, these limitations affect the size, not the significance, of the relationship. Another limitation of the model, however, is its lack of sensitivity to economic conditions.

For many years, motor vehicle fatalities have increased during prosperous economic times and decreased during recessions (*70*). The industrial production index tracks the number of highway fatalities reasonably well. Other measures of economic conditions, such as retail sales, correlate with monthly motor vehicle fatalities as well as measures of total driving (*71*). As noted in the analysis that follows, however, even when these variations in economic conditions, as measured by changes in employment, are accounted for, motor vehicle fatalities still appear to be significantly affected by the speed limit (*72*).

In a recent analysis a remarkable correlation was demonstrated between the number of motor vehicle fatalities and economic variables, yet a significant relationship continues to be indicated between fatalities and the speed limit (*72*). Using annual data from 1960 to 1982, this model accounted for economic fluctations by measuring changes in employment, unemployment, and labor force participation. The speed limit and the fuel shortage are accounted for with binary variables.

The model takes the form of:

$$F = H + b_1U + b_2E + b_3N + b_4A + b_5S$$

where

F = the annual number of motor vehicle fatalities,
H = constant,
U = unemployed workers (in thousands),
E = employed workers (in thousands),
N = people who are out of the labor force (in thousands),
A = a binary variable for the Arab oil embargo (1960 to 1973 and 1975 to 1980 = 0),
S = a binary variable for the 55 mph speed limit (1960 to 1973 = 0 and 1974 to 1980 = 1), and
b_1 to b_5 = parameters to be estimated.

Calibration yielded the following equation:

$$\text{Fatalities} = 96{,}781 - 1.8569U + 0.4971E + 0.9616N - 3995A - 4{,}824S.$$

These variables account for 98 percent of the variation in annual fatalities. Given the wide variation in the dependent variable, the model fits remarkably well (Figure 16).

By measuring economic changes through labor force variables, this model links economic changes to the population as a whole. When unemployment is low, indicating a booming economy, discretionary income is relatively high. Under such circumstances more travel of all types, particularly recreational travel, occurs. With more motorists on the road, the risks of fatalities increase.

These relationships are quite plausible. Total travel and total fatalities are highly correlated, because when driving increases, the exposure to risk increases. Despite the close fit between annual fatalities and labor force variables, the model also demonstrates a significant relationship between the decline in fatalities and the speed limit. The reduced number of fatalities associated with the speed limit is in the range of 3,700 to 5,900 lives annually. This estimate falls within the range of estimates from other approaches.

Mean Speeds and Speed Variance: A Lingering Uncertainty

In determining the extent to which the 55 mph speed limit affects safety today and in assessing the effects of future speed limit policy, the key

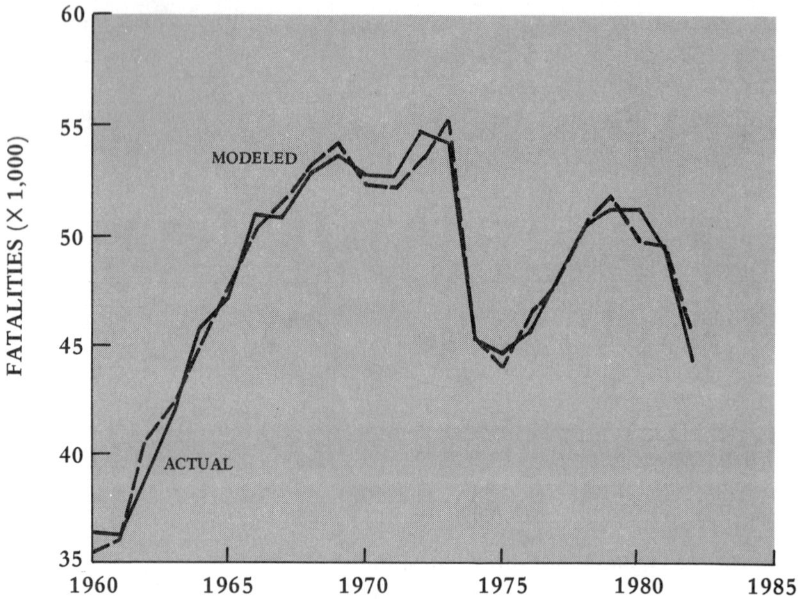

FIGURE 16 Fit of fatality model, 1960 to 1982, adjusting for the oil shortage and the 55 mph speed limit (*72*).

lies in how speed limits will affect the distribution of actual driving speeds. Two aspects of this distribution—the mean and the variance—have repeatedly surfaced in this analysis, and there are physical reasons why each of these characteristics affect highway safety. The variance of speed reflects differences of speed within the traffic stream. High variance may reflect a significant amount of lane changing, passing, or stop-and-go conditions. The probability of an accident occurring has been shown to be higher when speed variance is high. The mean speed, as noted earlier, affects stopping distances and the distance a motorist travels during any fixed reaction time. Further, the severity of accidents has been shown to be greater when speeds are high.

Fatality rates on state highway systems correlate positively with some measures of highway speeds. These correlations are statistically significant but not strong. Some measures of speed, such as the percent of traffic exceeding 55 mph and the 85th percentile speed, do not correlate with fatality rates. Others, such as the percentage of traffic exceeding 65 mph, correlate positively, though weakly, with the fatality rate (Table 8).

Data limitations may account for the weak correlation between these

TABLE 8 Correlation Between Speed Variables and Fatality Rates by State Highway System

Highway System	Percent Exceeding 55 mph	Percent Exceeding 65 mph	Number of States
Statewide	−0.02	0.27[a]	47
Interstates	−0.01	0.33[a]	48
Rural arterials	0.02	0.25[a]	45
Rural collectors	0.01	−0.09	37
Urban freeways	−0.35	−0.42	14
Urban arterials	−0.12	−0.08	21

[a]Correlation significant to 0.05.

variables. Because only about 3,500 fatalities occur on Interstate routes each year, random events can distort the fatality rate. One accident—a four-car pile-up, for example—would double the number of fatalities on the rural Interstate in some states.

Statistically, the importance of the speed variance can be illustrated by using a multiple regression approach. Highway safety characteristics such as traffic density, speed variance, percentage of travel exceeding 65 mph, enforcement activity, and the percentage of teenagers in the population can be explicitly included. The model is specified as follows:

$$\text{Fatality rate} = a + b_1X_1 + b_2X_2 + b_3X_3 + b_4X_4 + b_5X_5 + E$$

where

a = constant,
X_1 = traffic density,
X_2 = speed variance,
X_3 = percentage teenagers,
X_4 = citations per million vehicle miles,
X_5 = percent of state VMT on the Interstate, and
E = error term.

The regression explains about 60 percent of the variation among states and is statistically significant at the 0.01 level (Table 9). The unexplained variation could be a function of within-state variability and the inability to control for important safety characteristics such as geometric design of highways and topographical differences. Some unexplained variation in a cross-sectional model should be expected, particularly because of the near random nature of highway accidents.

Speed variance in this model is positively related to fatality rates; that is, states with wider variances in speed tend to have higher fatality rates.

Previous research has revealed wider variations in traffic speeds on individual highway segments to be related to a higher probability of accident occurrence (*6,7*). The 55 mph speed limit reduced both the average speed and the variations in speeds in 1974 by radically curtailing the amount of speeding above 65 to 70 mph. These facts suggest that the lower variances brought about by the 55 mph speed limit have been one of the reasons that the speed limit has reduced highway fatalities.

Indeed, some analyses suggest that the reduction in speed variance instead of the reduction in average speeds is the reason that reduced speed limits have improved highway safety (see Appendix C). Although available data are too sketchy to verify this point conclusively, they suggest that if the average speed of the traffic stream could be increased without increasing the variance of the speed, then the adverse effects on safety might be comparatively small. Although this poses an interesting set of questions for future research, it has little bearing on the conclusions reached in this chapter: whether based on its impact on mean speeds or its impact on speed variance, the 55 mph speed limit has substantially reduced the number and rate of highway fatalities. In setting future policy, the exact mechanism by which speed limits affect safety clearly needs to be better understood, and the possibility that speed variance may be a key consideration in this respect may help to select or design highways where high-speed driving results in the least deterioration of highway safety.

TABLE 9 Model of State Fatality Rates on Highways Posted at 55 mph

Explanatory Variables	Beta	Standardized Beta[a]	Standard Error	*t* Statistic
Traffic density	−0.0003	−0.3453	0.0001	3.09[b]
Speed variance	0.3712	0.4475	0.0772	4.81[b]
Percent teenagers	0.1721	0.2334	0.0785	2.19[c]
Citation rate	−0.0508	0.3562	0.0154	3.29[b]
Percent VMT on Interstates	−0.0183	−0.2798	0.0080	2.26[c]

NOTE: N = 48; R^2 = 0.64; mean squared error = 0.39; F ratio = 15.05 (significant at 0.01).

[a]Beta coefficients have been standardized to have unit variance; that is, the standard deviations have been set equal to 1. This allows comparison of the relative effect of the independent variables on the dependent variable.

[b]Significant at 0.01.

[c]Significant at 0.05.

At this point, however, the extent to which speed variance affects safety independently of the mean speed remains unclear, and the extent to which motorists can increase their mean speed without increasing the variance of speed has not been demonstrated (see Appendix H). Even if future research were able to isolate the effects of speed variance separately from mean speed, and even if ways were found to control the variance independently of the mean, there is still no reason to believe that increases in the mean speed, alone, would have no effect on safety: both physics and statistical analysis of actual experience indicate that average speeds have a bearing on safety. In summary, although future research on different measures of the speed distribution may point to highway design and traffic control improvements that minimize the effect of average speed on safety, the facts at this stage clearly demonstrate that the imposition of the 55 mph speed limit, through its impact on the distribution of travel speeds, caused a major reduction in highway fatalities, and that a reversal of this policy would cause an erosion of those safety benefits. The magnitude of this erosion could be smaller than the estimates presented here if future research reveals that speed variance is an independently controllable, key element in speed-limit policy, and if acceptable policy measures can be enacted to achieve such independent control of speed variance.

CONCLUSION

In the decade since the Arab oil embargo, the 55 mph speed limit, along with improved vehicles, highways, and other safety policies, has contributed to a safer driving environment. The reduced speeds and reduced speed variance of 1974 appear to be key factors in the dramatic decline in highway fatalities in 1974. Nationwide statistics and more detailed analyses of state experience indicate that 3,000 to 5,000 lives were saved annually in the early years of the 55 mph speed limit. International experience with speed limits is consistent with this estimate.

Some of the safety gains associated with the 55 mph speed limit may have diminished during the decade it has been in effect. Policies that rely on changes in motorist behavior tend to be most effective when they are first introduced, because at that time public awareness is high. The 55 mph speed limit is no exception, as reflected in average highway speeds that are gradually increasing from year to year. The effectiveness of the 55 mph speed limit may diminish over time, as motorist behavior reverts to former patterns. Nevertheless, average highway speeds remain below their 1973 levels. Furthermore, the distribution of driving speeds remains far less dispersed than it was in 1973. Because the variance in speeds continues to remain less dispersed than it was in 1973, the safety consequences of the 55 mph speed limit have not been eroded.

Today, more than a decade following the inception of the 55 mph speed limit, numerous changes have reshaped driving conditions, possibly modifying the extent to which fatalities would be increased if the speed limit were increased. These changes include improved emergency medical services, crackdowns on alcohol-impaired driving, physical improvements to highways, introduction of more stringent motor vehicle standards, increased use of smaller cars, increased volume of truck traffic, fewer teenage drivers, and more elderly drivers. Together such factors have yielded a net reduction in the nation's highway fatality rate, which is down from 3.59 deaths per 100 million vehicle miles in 1974 to 2.76 in 1983, a reduction of 23 percent.

The benefits of the 55 mph speed limit may be somewhat offset by other safety improvements. These improvements, however, only partly substitute for the benefits of reduced speed. Many of the physical relationships that make speed risky continue fundamentally unchanged. The distance traveled during perception times at high speeds are necessarily longer than at low speeds. The amount of vehicle deceleration during impact still increases disproportionately with speed. Crashes at higher speeds still have a greater probability of resulting in a fatality, a problem more pronounced with smaller cars.

Nevertheless, as improvements have been made to highways, vehicles, and medical services, the risk associated with higher speed driving has been reduced somewhat. Accordingly, the number of lives being saved as a result of the 55 mph speed limit might be smaller today than when the speed limit was first enacted. If this reduction is proportional to the reduction in the fatality rate overall (and accounting for increased travel and reduced compliance), then the number of lives saved per year as a result of the 55 mph speed limit currently would decrease to approximately 2,000 to 4,000.

Some verification that these safety benefits are still being realized can be derived from statistical analysis of speeds, fatalities, and other key factors during the past decade. Time series models are particularly useful in explaining the variation in highway deaths during the last decade, and these models demonstrate the continued importance of the 55 mph speed limit. Cross-sectional analyses of state data indicate that speed variance continues to be an important predictor of fatality rates. Applications of these statistical techniques indicate that a few thousand lives continue to be saved each year as a result of the 55 mph speed limit.

Pending changes in the highway environment could increase the importance of the 55 mph speed limit in the future. The elderly are the fastest growing age group among drivers, and they are more susceptible to injury in accidents, particularly high-speed crashes. The mix of traffic will become more hazardous as the share of total highway travel accounted for by heavy trucks increases. Slower speeds reduce the severity of crashes between cars and trucks. There will also be changes that diminish the influence of speed on safety. Continued improvements will be made in vehicle crashworthiness, traffic operations, emergency medical services, road design, and other aspects of the highway driving environment. Although all of these changes may alter somewhat the importance of the 55 mph speed limit in achieving highway safety, this law has saved a few thousand lives per year throughout the past decade and promises to continue to have substantial safety consequences in the decade ahead.

REFERENCES

1. National Safety Council. *Accident Facts*. 1982.
2. L.I. Griffin, III. "Analysis of the Benefits Derived from Certain Presently Existing Motor Vehicle Safety Devices: A Review of the Literature," Highway Safety Research Center, University of North Carolina. 1973.
3. *Fatal and Injury Accident Rates on Federal Aid and other Highway Systems*. Annual issues. Table FR-1. FHWA, U.S. Department of Transportation, 1968-1977.
4. *Highway Statistics*. Annual issues, 1973–1974. Table VS-1. U.S. Department of Transportation.
5. American Association of State Highway Officials. *A Policy on Geometric Design of Rural Highways*. Washington, D.C., 1965.
6. W. Haddon and S. Baker. "Injury Control." Chapter 8. *Preventive and Community Medicine*. Edited by C. Clark and B. Macmahon. Little, Brown and Co., New York, 1981.
7. R. Synder. "Human Impact Tolerance" *1970 SAE Transactions*, Vol. 79, New York, 1970.
8. J. O'Day and J. Flora. Alternative Measures of Restraint System Effectiveness: Interaction with crash Severity Factors. SAE Technical Paper 820798. Society of Automotive Engineers, Warrandale, Pa., 1982.
9. D. Solomon. *Accidents on Main Rural Highways Related to Speed, Driver, and Vehicle*. FHWA, U.S. Department of Transportation, 1964.
10. J.A. Cirillo. "Interstate System Accident Research—Study II—Interim Report II." *Public Roads*, Vol. 35., No. 3., 1968.
11. E. Hauer. "Accidents, Overtaking and Speed Control." *Accident Analysis and Prevention*, Vol. 3 pp. 1-13 Pergamon Press, 1971. Great Britain.
12. American Association of State Highway and Transportation Officials. *Effects of the 55 mph Speed Limit*. Washington, D.C., 1974.
13. R.F. Heckard, J.S. Pachuta, and F.A. Haight. *Safety Aspects of the National 55 mph speed Limit*. FHWA RD 76-191. FHWA, U.S. Department of Transportation, 1976.
14. E. Cerrelli. *Estimating the Safety Effects of the 55 mph Speed Limit*. DOT HS 802-475. NHTSA, U.S. Department of Transportation, 1977.

15. American Association of State Highway and Transportation Officials. *Safety Impact of the 55 mph Speed Limit*. Washington, D.C., 1977.
16. P. Johnson, T. Klein, P. Levy, and D. Maxwell. *The Effectiveness of the 55 mph National Maximum Speed Limit as a Life Saving Benefit*. DOT HS 805-811. NHTSA, U.S. Department of Transportation, 1980.
17. *Fatal and Injury Accident Rates on Federal Aid and Other Highway System*. Annual issues, 1973 and 1974, Tables FT1, FT31, FT3 and FT5. U.S. Department of Transportation.
18. International Bridge, Tunnel and Turnpike Association. Turnpike Fatality Rates, 1954-1981. Unpublished. Washington, D.C., 1984.
19. K. Cambell, R. Scott, and S. Tolkin. *Highway Safety Effects of the Energy Crisis on U.S. Toll Roads*. Highway Safety Research Institute, University of Michigan, Ann Arbor, 1976.
20. National Safety Council. *Accident Facts*. Annual issues, 1974-1977.
21. *Public Use of the National Parks: A Statistical Report, 1971-1981*. U.S. Department of the Interior.
22. *Census of Transportation: Travel During 1972*. Tables 3A and 3B. Bureau of the Census, U.S. Department of Commerce, Sept. 1973.
23. F. Council, L. Pitts, M. Sadorf, and O. Dart. *An Examination of the Effects of the 55 mph Speed Limit on North Carolina Accidents*. Highway Safety Research Center, University of North Carolina, Chapel Hill, 1975.
24. J. O'Day, J. Minihan, D. Golomb. *Study of the Effects of the Energy Crisis and the 55 mph Speed Limit in Michigan*. Highway Safety Research Institute, University of Michigan, Ann Arbor, 1975.
25. B.E. Burritt, A. Moghrabi, and J. Matthias. "Analysis of the Relation of Accidents and the 88-km/h (55 mph) Speed Limit on Arizona Highways." In *Transportation Research Record 609*, TRB, National Research Council, Washington, D.C., 1976.
26. W. Pudinski. *Report on Accident Reduction Variables*. California Highway Patrol. Sacramento, Calif., 1974.
27. B.Y. Chu and G.E. Nunn. "An Analysis of the Decline in California Traffic Fatalities During the Energy Crisis." *Accident Analysis and Prevention*, Vol. 8, 1976, pp. 145-150.
28. H.S. Dawson, Jr. "Analysis of Fatal Accident Trends on Maryland Highways 1970-1976." *Public Roads*, Sept. 1979.
29. P. Johnson, T. Klein, P. Levy. *Effect of the 55 mph Speed Limit on Fatal Crashes in Texas*. NHTSA, U.S. Department of Transportation, 1978.
30. T. Klein. *The Effect of the 55 mph Speed Limit on Traffic Accidents in Illinois*. DOT HS 805 400. NHTSA, U.S. Department of Transportation, 1980.
31. J. Wiorkowski and R.F. Heckard. "The Use of Time Series Analysis and Intervention Analysis to Assess the Effects of External Factors on Traffic Indices: A Case Study of the Effect of the Speed Limit Reduction and Energy Crisis in the State of Texas." *Accident Analysis and Prevention*, Vol. 9, 1977, pp. 229-247.
32. H. Golomb and J. O'Day. "An AID analysis of Texas Traffic Accident Data Before and During the Energy Crisis." *HIT LAB Reports*, Vol. 5, No. 7. Highway Safety Research Institute, University of Michigan, March 1975.
33. T.M. Borg. *Evaluation of the 55 mph Speed Limit*. Purdue and Indiana

State Highway Commission Joint Highway Research Program, West Lafayette, 1975.

34. K.R. Agent, D.R. Herd, R.G. Rizenbergs. "First Year Effects of the Energy Crisis on Rural Highway Traffic In Kentucky." In *Transportation Research Record 567*. TRB, National Research Council, Washington, D.C., 1976.
35. N. Enustun, D. Hornbeck, S. Lingeman, A. Yang. *Safety Aspects of the 55 mph Speed Limit*. TSD 295-74. Michigan Department of Transportation, Lansing, 1974.
36. W.D. Labrum. "The 55 mph Speed Limit and Fatality Reduction in Utah." *Traffic Engineering*, Sept. 1976.
37. Organisation for Economic Cooperation and Development (OECD) *Speed Limits Outside Built-Up Areas*. Paris, 1972.
38. G. Nilsson. *Trials with Differentiated Speed Limits During the Years 1968-72*. Report 177A. National Road and Traffic Research Institute, Linköping, Sweden, 1977.
39. G. Nilsson. *The Effects of Speed Limits on Traffic Accidents in Sweden*. Report 68. National Road and Traffic Research Institute, Linköping, Sweden, 1982.
40. M. Salusjärvi. *The Speed Limit Experiment on Public Roads in Finland*. Publication 7/1981. Technical Research Centre of Finland, Espoo, Finland, 1981.
41. R. Fieldwick. *The Relationship between Rural Speed Limit and Accident Rate*. Technical Report RF/81. National Institute for Transport and Road Research, South Africa, 1981.
42. European Conference of Ministers of Transport. *Costs and Benefits of General Speed Limits (Maximum and Minimum Speeds)*. Round Table 37, Paris, 1978.
43. W.J. Frith and J.B. Toomath. "The New Zealand Open Road Speed Limit." *Accident Analysis and Prevention*, Vol. 14, No. 3, 1982.
44. J.A. Pierce. "Safety Benefits of the Seat Belt Legislation and Speed Limit Reduction in Ontario." *Proc., 23rd Annual Conference, American Association of Automotive Medicine*, Morton Grove, Ill., 1977.
45. P. Scott and A.J. Barton. *The Effects on Road Accident Rates of the Fuel Shortage of November 1973 and Consequent Legislation*. Supplementary Report 236. Transport and Road Research Laboratory, Crowthrone Berksire, 1976.
46. R.A. Daltrey and D. Healy. "Change in Victorian Accident Frequency with the Introduction of the 100 km/h Absolute Speed Limit in 1974." *Proc., Australian Road Research Board*, Vol. 10, Part 4, 1980.
47. J. Christensen. "The Effects of General Speed Limits on Driving Speeds and Accidents in Denmark." *Proc., The Effects of Speed Limits on Traffic Accidents and Transport Energy Use*, Organisation for Economic Cooperation and Development. An Foras Forbartha. Dublin, 1981.
48. S. Godwin. "International Experience with Speed Limits During and Prior to the Energy Crisis of 1973-4." *Transportation Planning and Technology*. Vol. 9, No. 1, 1984, pp. 25-37.
49. T. Zieziulewicz and W. Tarrants. *Effectiveness and Efficiencies in Emergency Medical Services*. DOT HS 806 143. NHTSA, U.S. Department of Transportation, 1982.
50. K. Ham, M. Yoder, and K. Ramaswany. Emergency Medical System Eval-

uation: A Case Study of York County Pennsylvania. Pennsylvania Department of Health, Harrisburg, Pa., June 1980.
51. J. Waters. The Jacksonville Emergency Medical System, A Model for the Seventies. Department of Public Safety, Jacksonville, Fla., Undated.
52. D. Trunkey. "Trauma." *Scientific American*, Vol. 249, No. 2, Aug. 1983.
53. H.L. Ross. *Deterring the Drinking Driver: Legal Policy and Social Control.* D.C. Heath and Co., Lexington, Mass., 1982.
54. A.C. Wagenaar, *Alcohol, Young Drivers, and Traffic Accidents: Effects of Minimum Age Drinking Laws*, D.C. Heath and Co., Lexington, Mass., 1983.
55. C. Kahane. The National Highway Traffic Safety Administration's Evaluations of Federal Motor Vehicle Safety Standards. SAE Technical Paper 840902. Warrendale, Pa., May 1984.
56. H. Joksch. Evaluation of Motor Vehicle Safety Standards. Report No. DOT HS-801 012. National Technical Information Services, 1973.
57. A McLean. *Collection and Analysis of Collision Data for Determining the Effectiveness of Some Vehicle Systems*. Motor Vehicle Manufacturers Association, 1974.
58. Effectiveness, Benefits and Costs of Federal Safety Standards for the Protection of Passenger Car Occupants. U.S. General Accounting Office, 1976.
59. J.D. Graham and S. Garber. Evaluating the Effects of Automobile Safety Regulation. *Journal of Policy Analysis and Management*. Vol. 3, No. 2., 1984, pp. 206-224.
60. Small Car Safety: An Issue that needs Further Evaluation. CED 82-29. U.S. Government Accounting Office, 1982.
61. L. Evans. Car Mass and Likelihood of Occupant Fatality. SAE Technical Paper 820807. Warrendale, Pa. June 1982.
62. W. Haddon. "Small Car Deaths, Injuries Worst: Models Vary Greatly." *Status Report*, Vol. 17, No. 20, Insurance Institute for Highway Safety, Washington, D.C., 1982.
63. A. Malliaris, R. Nicholson, J. Hedlund, and S. Scheiner. Problems in Crash Avoidance and in Crash Avoidance Research. SAE Technical Paper 830560. Warrendale, Pa., March 1983.
64. L. Evans. Accident Involvement Rate and Car Size. Transportation Research Department, General Motors Research Laboratories, Research Publication GMR-4453. Warren, Mich., Aug. 1983.
65. L. Lohman and P. Waller. *Trucks: An Analysis of Accident Characteristics by Vehicle Weight*. Highway Safety Research Center, University of North Carolina, Chapel Hill, 1975.
66. J. Eicher, H. Robertson, G. Toth. Large Truck Accident Causation. Technical Report DOT HS 806 300. NHTSA, U.S. Department of Transportation, July 1982.
67. *Fatal Accident Reporting System, 1982*, Table 14. NHTSA, U.S. Department of Transportation, May 1984.
68. *Federal Highway Cost Allocation Study*. Final Report. FHWA, U.S. Department of Transportation, May 1982.
69. Abt Associates. Accident Trend Monitoring and Exploratory Analysis. *First Annual Highway Traffic Safety Trend Report*. Vol. II. Exhibit 2.34. NHTSA, U.S. Department of Transportation, April 1982.
70. D.F. Mela. *Review on Information on the Safety Effects of the 55 mph Speed*

Limit in the United States. NHTSA, U.S. Department of Transportation, 1977.

71. P. Hoxie, D. Skinner and G. Wang. *Socio-Economic Influence on Highway Fatalities*. Transportation Systems Center, Cambridge Mass., 1983.
72. S. Partyka. Simple Models of Fatality Trends Using Employment and Population Data. National Center for Statistics and Analysis. U.S. Department of Transportation, Aug. 1983.

NOTES

1. Because of reclassification of the federal-aid highway system in 1976, trends for primary (not including Interstates), secondary, and urban highways are inconsistent past 1976. Only the Interstates have a consistent trend, and the data extend from 1967 to the present.
2. These comparisons assume linear deceleration of 15 ft per second per second and reaction times by all drivers of 0.75 seconds.
3. This calculation assumes that fatality rates and reduced travel are linearly related. Although this assumption is unlikely to hold when travel is substantially reduced, under conditions of relatively small change (in this case 1.45 percent) the assumption of a linear relationship is quite safe.
4. Studies of speed limits and safety in Finland and Sweden rarely included a large enough number of fatalities to estimate the effects of reduced and more uniform speeds on fatality rates.
5. Although there is no precise technique for estimating the current life saving benefit of the 55 mph speed limit, the number of lives saved can be approximated by accounting for three major factors that have changed since 1974: total vehicle travel, the fatality rate, and vehicle speeds. Travel increased 24 percent between 1974 and 1983. The fatality rate has declined 23 percent during this same period, or to 0.77 of its previous value. Average speeds have increased about 40 percent of the amount they declined in 1974, but the variability in speeds appears to have remained constant. Given the interrelationship between these two speed measures and their importance to highway safety, this adjustment assumes that 20 percent of the benefit of the speed limit has been lost (the midpoint between a 40 percent loss of the decline in average speeds and no change in the variability in speeds). Based on the estimate of 3,000 to 5,000 lives saved in 1974, the adjustment for increased travel, improved safety, and increased speed can be applied to this base to estimate the lives saved in 1983 by assuming that each factor had a value of 1 in 1974. Thus travel in 1983 would equal 1.24 percent of its 1974 value; safety, as measured by the fatality rate, would equal 0.77 percent of its previous value, and speed would equal 0.80 of its previous value. Multiplying these adjustments by the estimate of 3,000 to 5,000 lives saved in 1974 results in an estimate of lives saved in 1983 of from 2,259 to 3,730, or rounded off to 2,000 to 4,000.

4
Estimated Reductions in Injuries and Injury Severity

Although severe data limitations make firm conclusions impossible, the 55 mph speed limit appears to have reduced the number of highway injuries substantially. Between 1973 and 1974, 182,400 fewer persons were injured in highway accidents (*1*). This represents a decline of 6.4 percent; on rural Interstate routes, where the 55 mph speed limit had the most effect on speeds, injuries declined by 31 percent between 1973 and 1974. More than one-half of the reduction in injuries occurred on the highway systems most directly affected by the 55 mph speed limit (Interstates and other rural primary and rural state highways).

Assuming that the reduction in injuries on high-speed roads is proportional to the reduction in fatalities on these same roads, between 51,000 and 82,000 fewer injuries occurred in 1974 as a result of the 55 mph speed limit. About 3,500 to 5,700 of these reduced injuries would have been serious, severe, or critical. As a result of the reduced number of injuries, total expenditures on medical bills, legal and court costs, and motor vehicle repair costs were lowered by between $122 to $240 million (in 1983 dollars). More important than these dollar savings, however, fewer and less severe injuries reduce the pain and suffering caused by traffic accidents.

INJURY DATA LIMITATIONS

The effect of slower driving speeds on accidents is clouded by uncertainty about the number of accidents and injuries averted because of the 55 mph speed limit. Estimating the effect of the speed limit on injuries and accidents is complicated by many of the same confounding influences that make it difficult to estimate reduced fatalities. In addition, injury data have many reporting problems that make these data difficult to work with.

Data on highway injuries have been less carefully collected and recorded than data on fatalities over the years for a variety of reasons. Most injuries are minor and many are not reported. With millions of injuries from motor vehicle accidents occurring each year, a census of injuries with the same level of detail available for fatalities would be extremely expensive to take. Some states collect aggregate injury information, but trend data are unavailable in consistent classifications. The U.S. Department of Transportation has recently developed the National Accident Sampling System to provide estimates of injuries and injury severity based on a representative sample, but these data have only been available since 1979. Injuries declined in 1974, but the wide variety of injuries, together with the relatively imprecise classification of injuries relied on at the time, makes it difficult to isolate the types of injuries most likely to result from reduced high-speed driving.

In the mid-1970s most highway patrols used a similar scale, known as the Police Injury Code, for measuring injury severity. This scale classified injuries with visible signs of seriousness, such as bleeding wounds or dislocated members, as "A" injuries. Other signs of injury, such as bruises, abrasions or swelling, were classified as "B" injuries. Victims who complained of pain or momentary unconsciousness, but showed no visible signs of injury, were classified as "C" injuries. This obviously rough delineation of injury severity cannot be consolidated across states. A sample of police ratings of injury severity in 17 states revealed that the proportion of all injuries classified as A level ranged from 13 to 65 percent (*3*). Some discrepancy across states would be expected, but a 5 to 1 ratio is unlikely. In addition, this classification is quite broad. Injuries might in actuality be less serious but still might be classified alike. In recent years a more refined injury classification—the Abbreviated Injury Scale (AIS)—has been developed (Table 10). This scale classifies survivable injuries on a five-point scale, with AIS 5 the maximum survivable injury. Estimates of injuries resulting from motor vehicle accidents at the national level have been available since the development

TABLE 10 Representative Motor Vehicle Injuries by Abbreviated Injury Scale Level (*2*)

AIS Code	Injury Severity Level	Representative Injuries
1	Minor injury	Superficial abrasion or laceration of skin, digit sprain, first-degree burn, head trauma with headache or dizziness (no other neurological signs).
2	Moderate injury	Major abrasion or laceration of skin, cerebral concussion (unconscious less than 15 minutes), finger or toe crush/amputation, closed pelvic fracture with or without dislocation.
3	Serious injury	Major nerve laceration; multiple rib fracture (but without flail chest); abdominal organ contusion; hand, foot, or arm crush/ amputation.
4	Severe injury	Spleen rupture, leg crush, chestwall perforation, cerebral concussion with other neurological signs (unconscious less than 24 hours).
5	Critical injury	Spinal cord injury (with cord transection), extensive second- or third-degree burns, cerebral concussion with severe neurological signs (unconscious more than 24 hours).

of the National Accident Sampling System in 1979 (*4*), unfortunately not until well after the introduction of the 55 mph speed limit.

INJURY RATE TRENDS

Nationwide Trends in the United States

Injury rates (injuries per 100 million vehicle miles traveled) declined in 1974 on the highway systems with reductions in posted speed limits. In 1974 injury rates on rural Interstates declined 29 percent below the injury rates of 1973 (Figure 17). The reductions on other rural primary highways were less pronounced (12 percent below 1973 levels) but still well below the previous year. Local rural roads and highways showed little change.

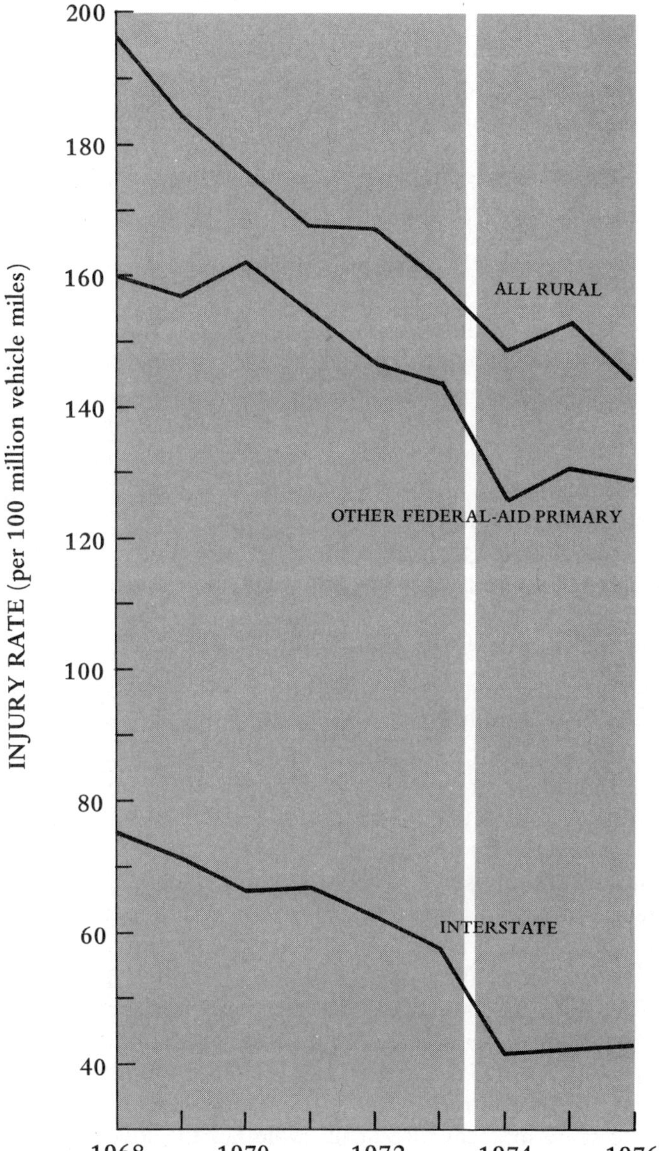

FIGURE 17 Injury rates on rural highways, 1968–1976 (*1*).

Injury rates on urban highways reveal a similar pattern (Figure 18). Injury rates on urban Interstates declined 20 percent below the 1973 rate. The injury rates for all urban highways and urban primary highways—systems posted at or below 55 mph to begin with—actually increased slightly in 1974.

Injury rates on Interstate highways and rural primary highways appear to have been affected by the reduced speed limit in 1974 because the trends continued below the previous levels through 1976, after the energy crisis had abated.

Injury Trends

The effect of the speed limit policy on injury rates can also be demonstrated with state-level injury data and more sophisticated trend analysis. Some states have monthly injury rate data that increase the number of observations sufficiently to apply time series modeling techniques. Monthly injury rates in Texas for the period 1968 to 1975 indicate a statistically significant relationship with binary variables measuring the energy crisis and the speed limit policy (*5*). The trend is also significantly reduced when regressed against the speed limit variable alone. Similar results have been demonstrated with monthly injury data from Illinois (*6*).

International Experience with Speeds and Resulting Injuries

Several other nations also have examined how speed limits affected injuries as well as fatalities. Differences in classification and reporting practices obviously limit international comparisons, but, in general, all the nations that lowered their speed limits recorded pronounced reductions in injuries and injury accidents, or both. (See the discussion of international experience in Chapter 3.)

International comparisons, state studies, and national data indicate a reduction in injury rates stemming from the 55 mph speed limit, but injury rate data do not allow for estimating reductions in the various types of injury. Thus, they do not indicate whether relatively minor injuries, such as bruises and abrasions, declined more or less than more serious injuries involving crushed bones and ruptured organs. Not only are severe injuries (AIS 3–5) the cause of more human suffering and disability, they are also more costly to individuals and society because they require more medical care. Thus, whether the speed limit reduces these more serious types of injuries is of particular interest. Although national data for the mid-1970s are too aggregate to permit analysis of the speed limit's effect on specific types of injuries, some states have collected detailed data that permit this exploration, as is discussed next.

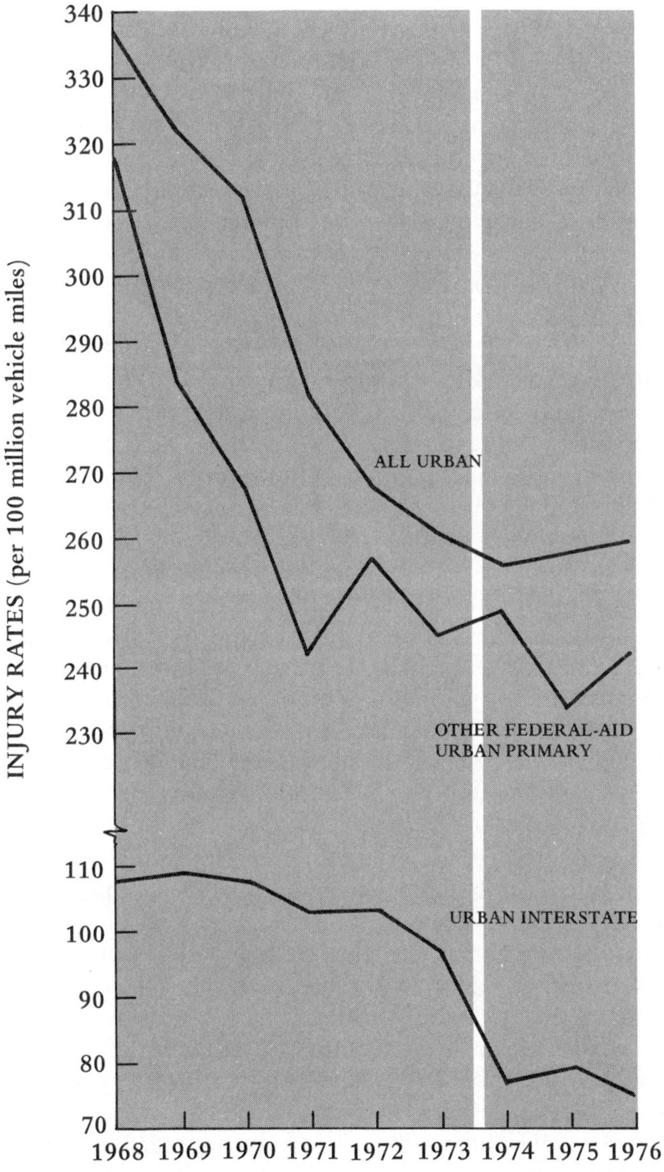

FIGURE 18 Injury rates on urban highways, 1968–1976 (*1*).

State-Specific Data on Injury Severity

Before examining actual data on the effect of the 55 mph speed limit by type of injury, it should be noted that there are a wide range of plausible hypotheses about how speed might affect the distribution of injuries. At one extreme, the reductions in speed at the time of impact could be just sufficient to reduce fatalities, but not sufficient to reduce severe or serious injuries to any great degree. In this instance, the net effect would be an increase in severe injuries because accidents that otherwise would have produced fatalities would now yield severely injured survivors. At the other extreme, all injuries might be reduced by the same proportion. Crashes that might have resulted in a fatality at higher impact speed would instead result in a critical injury, whereas crashes that might have resulted in a critical injury would instead result in a severe injury, and so forth. Another plausible hypothesis is that the 55 mph speed limit would reduce serious, severe, and critical injuries more than it would reduce moderate and minor injuries. This outcome would be similar to the observed reductions in injuries in studies of the effectiveness of safety belts. A review of the published studies of safety-belt effectiveness reveals that these safety devices reduce fatalities and the most serious injuries more than the less serious ones (*7*). By analogy, it is plausible that reduced crash severity because of slower impact speeds would have the most pronounced effect on the most serious injuries.

Among these three hypotheses of the effect of reduced speed on injury severity, little empirical evidence can be found to support the assumption that only fatalities have been reduced by the 55 mph speed limit. Reported data from North Carolina for 1973 and 1974 suggest that all types of injuries were uniformly reduced in 1974 on high-speed roads (*8*). These data are classified by type of injury (in the Police Injury Code discussed earlier) and they can be tabulated by posted speed limit and highway system. They are also reported for driver injury only. By limiting the analysis to driver injuries, the possible confounding effects of having more injuries per accident due to increased vehicle occupancy rates is eliminated.

During the first quarter of 1974 injuries of almost all types declined on all highways, even those posted below 55 mph (Table 11). This similarity may result from the modest speed changes that occurred in North Carolina. Few highways, aside from the Interstates were posted above 55 mph. On Interstate highways level A injuries (the most serious nonfatal injury rating) declined by 65 percent, far greater than the declines in levels B and C injuries (40 and 34 percent declines, respectively). In addition, accidents in which no injuries were reported declined sharply

TABLE 11 Driver Injuries in North Carolina, 1973 and 1974 (*8*)

	First Quarter			Second Quarter			Third Quarter		
Injury	1973	1974	Change (%)	1973	1974	Change (%)	1973	1974	Change (%)
Speed Limit Less Than 55 mph									
K	53	39	−26.4	51	46	−9.8	60	51	−15.0
A	717	576	−19.7	788	639	−18.9	739	627	−15.0
B	1,715	1,815	+5.8	2,252	2,097	−7.4	2,266	2,347	+3.5
C	2,463	2,329	−5.4	2,689	2,350	−12.6	2,553	2,737	+6.7
None	33,022	27,981	−15.3	32,134	29,188	−9.2	31,460	31,817	−1.1
Total	37,970	32,740	−13.8	38,112	34,320	−9.9	37,078	37,579	+1.3
Speed Limit Greater Than or Equal to 55 mph									
K	148	130	−12.2	188	127	−32.4	194	148	−23.7
A	962	742	−22.9	1,147	1,009	−12.0	1,238	1,066	−13.9
B	1,596	1,307	−18.1	1,989	1,869	−6.0	2,026	1,963	−3.1
C	1,337	1,069	−20.0	1,562	1,364	−12.7	1,548	1,446	−6.6
None	15,056	10,632	−29.4	15,461	13,064	−15.5	16,375	14,984	−8.5
Total	19,099	13,880	−27.3	20,347	17,433	−14.3	21,381	19,607	−8.3
Interstates									
K	14	15	—	11	11	—	10	5	—
A	83	29	−65.0	73	37	−49.3	77	47	−38.9
B	107	64	−40.2	140	85	−39.3	164	97	−40.1
C	110	73	−33.6	142	78	−45.1	135	83	−38.5
None	1,286	671	−47.8	1,280	758	−40.8	1,356	871	−35.7
Total	1,600	852	−46.7	1,646	969	−41.1	1,742	1,103	−36.7

on all highways posted at 55 mph. Data in the first quarter probably include the largest confounding influences of the fuel shortage (the Arab oil embargo ended in March 1974); thus data from later months in the year may well provide a better indication of the effect of the speed limit on injuries.

In the second two quarters of 1974 injuries in all classifications on the Interstates declined with some consistency. Levels A–C injuries declined between 39 and 49 percent in the second quarter and between 38 and 40 percent in the third quarter. In both quarters, accidents in which no injuries occurred declined by roughly the same proportion. Although fourth-quarter data were not reported in this study, they would be of less use because the energy crisis and voluntary speed reductions in late 1973 would have distorted the quarterly comparisons. Based on the second- and third-quarter data, the patterns on Interstate highways suggest that all types of injuries were fairly uniformly reduced by the 55 mph speed limit.

Injury severity data for Texas (provided by the State Highway Department) indicate large percentage reductions in all types of injuries on Interstate, US highways, and state highways (Table 12). The most serious, nonfatal injuries on these highways declined by 33 percent in 1974, which is slightly more than the decline in fatalities on Interstates (down 25 percent) and equal to the decline in fatalities on U.S. and state highways. Less serious injuries (B's and C's) did not decline as sharply, although they did decline by 19 to 25 percent.

Injury data from 23 states reporting to the National Safety Council in both 1973 and 1974 are less precise in the treatment of injury severity than the North Carolina and Texas data cited here, but a reduction in injuries is indicated in 1974, particularly in serious injuries (Table 13). Injuries with the most serious classifications (incapacitating) declined in about the same proportions as fatalities. Lesser injury types did not decline as much. Because these data include injuries on all types of roads and highways (urban and rural), these results cannot be as closely linked to the 55 mph speed limit as those from Texas and North Carolina, where the performance of high-speed roads could be isolated.

Two facts stand out from the preceding review of national and state experience with speed limit reductions and reduced injuries in 1974. First, a reduction in injuries accompanied the introduction of the 55 mph speed limit, particularly on the highway systems most directly affected by the speed restriction. Second, all types of injuries were reduced in 1974. There is some evidence to suggest that the effect on the most serious injuries (Police Injury Code A's) may have been somewhat larger

TABLE 12 Texas Rural Highway Accidents, 1973–1974

Highway System	Injury	1973	1974	Change (%)
Interstate	K	271	203	−25.1
	A	838	559	−33.3
	B	1,431	1,077	−24.7
	C	969	778	−19.7
U.S. and state	K	1,185	789	−33.4
	A	3,768	2,525	−33.0
	B	6,537	4,855	−25.7
	C	3,658	2,963	−19.0
Farm-to-market	K	341	353	+3.5
	A	1,477	1,371	−7.2
	B	2,674	2,502	−6.4
	C	1,417	1,348	−4.9
County roads	K	240	210	−12.5
	A	1,414	1,222	−13.5
	B	3,019	2,956	−2.1
	C	1,421	1,599	+12.5
Town	K	213	227	+6.6
	A	1,188	932	−21.5
	B	2,681	2,333	−13.0
	C	1,764	1,680	−4.8

than on less serious ones. However, the injury reductions on Interstates in Texas and North Carolina suggest that injuries of all types were reduced in an approximately uniform fashion on the highest speed highways.

Estimate of the Number of Injuries Reduced

At first glance, the simplest assumption in estimating the effect of the 55 mph speed limit on injuries would be to apply the same percentage reduction in total fatalities attributed to the 55 mph speed limit to the number of injuries. Such an estimate, however, would greatly overstate the effect of the 55 mph speed limit in reducing injuries. Whereas one-half of all fatalities occur on highways posted at 55 mph, the majority of injuries occur on low-speed roads in urban areas (Figure 19). For this reason, the appropriate base for deriving the estimate of the speed limit's impact on injuries is the reduced number of injuries on the highway systems most directly affected by the speed limit.

TABLE 13 Reductions in Total Statewide Injuries, 1973–1974

Injury Category	Percent Change 1973–1974
Fatal	–12.9
Incapacitating	–11.6
Evident	–7.8
Possible	–8.1

SOURCE: Data for 23 states provided by the National Safety Council.

Interstates and rural primary and rural state secondary highways had 102,500 fewer injuries and 6,300 fewer fatalities in 1974 (*1*). Some of these reductions are attributable to the energy crisis and the historic trend in improved safety. The assumption that the reduced injuries on these highways due to the 55 mph speed limit is in the same proportion as reduced fatalities provides a reasonable estimate of the effect of reduced speed on injuries. Because 3,000 to 5,000 fewer fatalities occurred in 1974 as a result of the 55 mph speed limit (Chapter 3), then 50 to 80 percent of this reduction is attributable to the 55 mph speed limit (3,000 to 5,000 divided by 6,300). Applying this same percentage to the reduced number of injuries on these same highways results in an estimate of 51,000 to 82,000 fewer injuries accounted for by the 55 mph speed limit.

Estimate of Reduced Injury Severity

In addition to the assumption that all types of injuries would be reduced uniformly, a further assumption is needed to estimate the distribution of injuries. Until the recent development of the National Accident Sampling System, national injury data were not available classified by type of injury. The estimates in this report assume that injuries in 1974 were distributed the same as the most recent estimate for seriously injured survivors (Table 14). This assumption is conservative when applied to 1973 and 1974 data. To the extent that safety has improved since 1974, injury severity would be expected to decline. Thus applying the 1982 injury distribution may classify fewer injuries as serious than actually were serious (calculations showing distributions are given in Appendix I, Tables A-24 and A-25). Applying the 50 and 80 percent reduction factor to injuries across the estimated change in all AIS categories between 1973 and 1974 results in the range of estimates given in Table 15.

These estimates suggest that the most serious injuries (AIS 3–5) declined

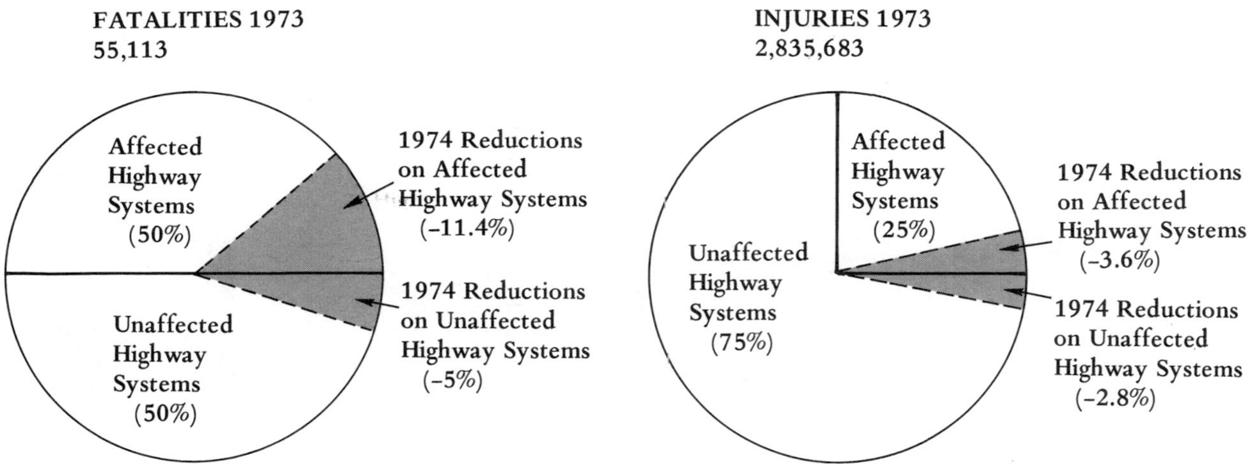

FIGURE 19 Fatalities and injuries, 1973 (*1*).

TABLE 14 Percentage Distribution of Known Injured Survivors by AIS Category

AIS	Urban (%)	Rural (%)
1	86.77	74.77
2	9.48	17.89
3	3.12	6.37
4	0.45	0.67
5	0.17	0.29
	100.00	100.00

SOURCE: Estimates provided by the National Center for Statistics and Analysis, National Highway Traffic Safety Administration, based on the National Accident Sampling System 1982 data.

by 3,500 to 5,700 in 1974 because of changes in driver behavior. This estimate may well be low because of the use of the 1982 distribution for injured survivors to estimate the injury distributions for 1973 and 1974. For this reason, the higher end of the range for serious injuries may be the better estimate of the effect of the 55 mph speed limit.

As noted in Chapter 3, overall highway safety has improved since the reduction in maximum speeds to 55 mph. With fewer fatalities due to other safety policies and demographic changes, the number of lives saved as a result of the 55 mph speed limit has declined somewhat. Applying the same reductions in lives saved to the reduced injury estimates suggests that these reductions could now be from 20 to 30 percent smaller. This would indicate an overall reduction in injuries for 1983 of 36,500 to 65,500. The current estimated impact of the 55 mph speed limit across all injury categories is given in Table 15. These estimates are used in the section on economic savings that follows.

TABLE 15 Estimated Reductions in Injuries by AIS Classification

	Estimated Reductions in Injuries	
AIS	1974	1983
1	40,000–62,500	28,000–50,000
2	8,500–13,800	6,000–11,000
3	3,100–5,000	2,200–4,000
4	300–500	210–400
5	100–200	70–160

ECONOMIC SAVINGS

The National Highway Traffic Safety Administration (NHTSA) estimated that Americans spent $3.3 billion for medical treatment of injuries sustained in traffic accidents in 1980 (*9*). Billions more were spent to repair or replace damaged automobiles and to cover legal and insurance costs. These expenses diverted resources from goods and services that would otherwise have been purchased. By reducing injuries and fatalities, the 55 mph speed limit reduces the unnecessary social costs of motor vehicle accidents, as outlined in the following sections.

Medical Costs

The medical expenses associated with motor vehicle injury accidents covers the costs of emergency medical services, including ambulance transportation from the scene of the accident, hospital costs for emergency room services, physician and surgeon expenses, costs for the initial hospital stay, and follow-up services. In extremely severe cases (AIS 5) there will be further costs associated with modifying the home so that the victim can cope with any disability that impairs normal functioning.

Although most such costs are generally borne in the first year after an accident, many occur in subsequent years as well. AIS 5 injuries typically require long periods of follow-up care, sometimes continuing for the remainder of the victim's (reduced) lifetime. In rare instances long-term rehabilitation costs can exceed $1 million (*10*). Because many of these expenses will be paid sometime in the future, they are discounted to reflect the present value of these outlays (*9,11*). (A conventionally applied discount rate of 10 percent is also assumed here.)

On the basis of the preceding estimates of the number of lives saved and the injuries averted as a result of slower driving, the nation's medical bill is reduced by $50 to $90 million each year. This reduction in medical costs represents between 1 and 2 percent of the nation's annual medical bill for treating those injured and killed in motor vehicle accidents.

Legal and Court Costs

In 1980 total insurance liability premiums as a result of motor vehicle accidents totaled $23.3 billion (*9*). On the basis of a number of studies conducted by the U.S. Department of Transportation and the insurance industry, the National Highway Traffic Safety Administration estimates that about 15 percent of total motor vehicle insurance premiums result from legal expenses due to injuries and fatalities suffered in motor vehicle accidents (*9*). On the basis of the average costs for legal expenses due to motor vehicle accidents, the 55 mph speed limit reduces legal costs by $20 to $55 million annually.

Lost Productivity

In addition to the incalculable pain and suffering that accidents impose on victims and their families, accidents also impair productivity. Those killed in car crashes lose a lifetime of productive activities. Those seriously injured are not productive while they recuperate and some will never be as productive after the injury or will not live as long because of the injuries sustained. To the extent that the 55 mph speed limit reduces the annual toll of death and injuries, it helps increase the nation's productivity. The size of this benefit is almost impossible to estimate because it depends on a host of uncertain factors, including whether and when an injured employee returns to work. Nor do these economic losses in any sense measure the individual grief and suffering associated with accidents. Nevertheless, allusion is made to these costs here because they constitute economic costs, some of which are reduced as a result of the 55 mph speed limit, but are not included in the computation of economic costs examined here.

Property Damage

One large and visible cost of highway accidents is the damage to the vehicles involved. Americans spend about $6 billion annually to repair vehicles involved in crashes (*4*,*9*). This estimate is based on reported property damage accidents and accidents involving injury or fatality. Many other accidents are never reported. The number of these accidents and the costs of repair are not well established; thus estimates of the total expenditures to repair motor vehicles are quite speculative. Based on the average costs of reported crashes, the 55 mph speed limit reduces property damage by about $52 to $95 million. This estimate is probably somewhat low because it is based only on the reduced number of injury accidents; however, there is little basis for speculating how much higher it would be if data on property-damage-only accidents and unreported accidents were available.

SUMMARY

The reduced and more uniform speeds brought about by the 55 mph speed limit coincided with a pronounced decline in injury rates on the

TABLE 16 Direct Costs of Motor Vehicle Accidents Reduced by the 55 mph Speed Limit

Category	Estimated Savings ($000s)
Medical	50,000–90,000
Legal	20,000–55,000
Property damage	52,000–95,000
Total	122,000–240,000

highway systems directly affected by lowered speed limits. These injury rates remained below pre-1974 levels after the fuel shortage abated. Generally, the pattern of injury reduction parallels that of fatality reduction, and the 55 mph speed limit appears to be a key factor accounting for the reductions.

About 52,000 to 82,000 fewer injuries in 1974 are attributable to the 55 mph speed limit. Of these, about 3,500 to 5,700 would have been serious, severe, or critical (according to the AIS classifications). Both of the preceding estimates are based on an assumption that the reductions in injuries caused by the speed limit on high-speed roads are proportional to the reduction in fatalities on those same roads.

In addition to avoiding some of the immense and immeasurable cost that traffic accidents impose in terms of human suffering, the 55 mph speed limit also reduces medical bills, legal costs, and the costs of motor vehicle damage. These costs are reduced by $122 to $240 million annually (Table 16).

REFERENCES

1. *Fatal and Injury Accident Rates on Federal Aid and Other Highway Systems, 1975*. FHWA, U.S. Department of Transportation, 1977.
2. N. Hartunian, C. Smart, and M. Thompson. *The Incidence and Economic Costs of Major Health Impairments*. Lexington Books, Mass.,1981.
3. R. E. Scott. "Evaluation of Severity Codes on Accident Data." *HIT Lab Reports*. Highway Safety Research Institute, University of Michigan, Ann Arbor, 1972.
4. *National Accident Sampling System, 1981*. NHTSA, U.S. Department of Transportation, 1983.
5. J. Wiorkowski and R. Heckard. "The Use of Time Series Analysis and Intervention Analysis to Assess the Effects of External Factors on Traffic Indices: A Case Study of the Effect of the Speed Limit Reduction and Energy Crisis in the State of Texas." *Accident Analysis and Prevention*, Vol. 9, 1977, pp. 229-247.

6. T. Klein. *The Effect of the 55 mph Speed Limit on Traffic Accidents in Illinois*. NHTSA, U.S. Department of Transportation, 1980.
7. B.J. Campbell and D.W. Reinfurt. The Degree of Benefit of Belts in Reducing Injury—An Attempt to Explain Study Discrepancies. SAE Technical Paper 790684. Society of Automotive Engineers, Warrendale, Pa., 1979.
8. F. Council, L. Pitts, M. Sadorf, and O. Dart, *An Examiniation of The Effects of the 55 mph Speed Limit on North Carolina Accidents*. Highway Safety Research Center, University of North Carolina at Chapel Hill, 1975.
9. *The Economic Cost to Society of Motor Vehicle Accidents*. NHTSA, U.S. Department of Transportation, 1983.
10. *Insurer Study of PIP Serious Injury Claims: Second Follow-Up, 1982*. All Industry Research Advisory Council, Oak Brook, Ill., Dec. 1982.
11. *Alternative Approaches to Accident Cost Concepts*. FHWA, U.S. Department of Transportation, 1984.

5

Taxpayer Costs and Benefits

Federal, state, and local governments spend more than $2 billion annually to compensate thousands of motor vehicle accident victims for medical costs or loss of earnings. Taxpayers also pay nearly $500 million in compensation and sick leave for public employees' lost time that results from accidental injuries on the highways. The Congress, aware of the magnitude of these costs, requested that this study highlight the extent to which these costs are reduced by the 55 mph speed limit.

Because of the enactment of the 55 mph speed limit, deaths and injuries on the nation's highways are reduced and the taxpayers are saved approximately $65 million annually. Reduced medical payments by programs such as Medicare and Medicaid are included in the overall reductions in medical costs given in Chapter 4. Additional enforcement is required to maintain compliance with the law, and this costs roughly $118 million annually. The revenue generated from fines, although not benefiting enforcement agencies directly, largely offsets the additional cost of enforcing the 55 mph speed limit.

The estimates provided in this chapter simply measure the impact of the 55 mph speed limit on the budgets of agencies and programs supported by general tax revenues. These estimates do not in any way address the major benefit of the 55 mph speed limit—the reduced pain and grief caused by highway accidents. Instead, they only indicate the influence of the speed limit on publicly supported programs, which ultimately affects the taxes paid by individuals.

The speed limit affects public budgets in three basic ways. Fewer serious injuries and fatalities reduce the demand for medical and social service assistance to the individuals eligible for these programs. With fewer disabling accidents, and fewer individuals out of work because of their injuries, the taxes these individuals pay continue to flow into public treasuries. Finally, the speed limit places additional demands on public budgets because of requirements for additional enforcement and for monitoring state compliance.

In order to estimate the benefits and costs of the 55 mph speed limit to taxpayers, some simplifying assumptions have been made. As noted in Chapters 3 and 4 of this report, the speed limit saves a few thousand lives each year, and probably reduces serious to severe injuries (AIS 3–5) on the order of 3,000 to 4,000 annually. These benefits vary from year to year and may have been greater at the time the speed limit was introduced than they are now. Throughout this chapter the estimates of the impact of the speed limit are based on the assumption that it reduces fatalities by 3,000 and serious injuries (AIS 3, 4, and 5) by 3,460 each year. These estimates have been used to provide a relative order of magnitude rather than a precise measure of impact.

In addition to the uncertainty about the precise number of reduced injuries and fatalties attributable to the 55 mph speed limit, the effect these reductions have on taxpayer-supported programs can only be approximated. Costs of major public programs such as the Old Age, Survivors, and Disability Insurance (OASDI) program, which has annual outlays of more than $100 billion, are not broken down by the type of expenditure. To estimate the impact of a single safety policy on this program requires assumptions about the number of people who have avoided injury, the number of people who would be eligible for this program, and the approximate claim they might make on the program budget. These assumptions and the calculations are provided in Appendix G and are summarized here. These estimates also assume that individuals in the population eligible for the public programs supporting accident victims have traffic accidents in proportion to their representation in the population. To the extent that lower-income persons drive less than the average American, these estimates may overstate the taxpayer costs associated with accidents. The data required to correct for any overestimation, however, are not available. In all cases the benefits and costs of the speed limit in tax dollars should be considered as the best available, though necessarily approximate, estimates.

TAXPAYER BENEFITS

The 55 mph speed limit reduces the cost of taxpayer-supported programs in four ways. Fewer injuries reduce the cost of medical assistance programs and fewer fatalities and disabling injuries reduce the number of families that need public assistance. Because fewer serious accidents and injuries are experienced by public employees, overall personnel costs are reduced. In addition, many state and local government costs are indirectly influenced by the number of serious accidents. Because of the impact of the 55 mph speed limit less demand is placed on coroner's offices, government insurance costs are reduced, and the amount of litigation involving state and local governments is reduced. Finally, because fewer individuals are killed or disabled, the taxes they pay to all levels of government are not interrupted. The savings to these programs, agency budgets, and public treasuries total approximately $65 million annually, as summarized in the following sections.

Savings to Publicly Supported Medical and Family Assistance Programs

Old Age, Survivors, and Disability Insurance

Almost one-half of the public costs due to motor vehicle accidents are borne by the Old Age, Survivors, and Disability Insurance program, which is administered by the Social Security Administration (Table 17). Almost all employed adult Americans are covered by this program, which is largely funded through payroll deductions. Total OASDI outlays in 1982, for all purposes, exceeded $150 billion (*2*). Of these total outlays, motor vehicle accidents of all types were responsible for approximately $1.5 billion (*1*). These costs result from lump-sum and survivor benefits to families of fatal accident victims and benefits to persons disabled in accidents. The largest single cost component results from survivor benefits, which can extend over many years to help support the families of fatal accident victims. The cost of survivor benefits alone would increase by almost $38 million dollars if 3,000 more persons were killed because of increased vehicle speeds (Table 17). Lump-sum awards to the families of fatal accident victims would increase by $800,000. Disability payments to individuals permanently injured in a motor vehicle accident would increase annually by $1.5 million. Overall, therefore, because of the 55 mph speed limit the cost of the OASDI program is reduced by about $40 million annually.

Although these savings are in the millions of dollars, and hardly reflect the benefits of reduced grief and suffering, the savings due to the 55 mph speed limit barely reduce the total benefits paid out by OASDI because of motor vehicle accidents. Assuming that the 55 mph speed limit saves 3,000 lives each year, the annual savings to OASDI represent about 3 percent of the total costs incurred by motor vehicle accidents.

TABLE 17 Direct Public Program Costs of Motor Vehicle Accidents (1983 dollars) (*1*)

Program	Total ($)	Savings Due to the 55 mph Speed Limit[a] ($)
Old Age, Survivors, and Disability Insurance (OASDI)	1,262,000,000	40,000,000
Medicare	142,000,000	5,000,000
Medicaid	640,000,000	7,000,000
Supplemental Security Income (SSI)	82,000,000	175,000
Aid to Families with Dependent Children (AFDC)	23,700,000	1,800,000
Food Stamps	3,300,000	60,000
Title XX	20,745,000	210,000
State Charity and Public Assistance	28,500,000	1,250,000
Public Employee Workmen's Compensation and Sick Leave	440,000,000	400,000
Total	2,642,245,000	55,660,000

[a] See Appendix G for estimates.

In terms of total fatalities, the 3,000 lives saved represents between 6 and 7 percent of annual accidental highway deaths. Most of the discrepancy between a 3 percent reduction in cost and a 6 to 7 percent reduction in fatalities can be explained by program eligibility—OASDI only covers working adults. But in a more general sense, the significant savings in cost due to the 55 mph speed limit only appear small when compared to enormous costs that result from the 45,000 deaths and 200,000 serious injuries (AIS 3–5) occurring each year on all roadways.

Medicaid

The largest program savings, other than those to the OASDI program, accrue to the Medicaid program. Medicaid is administered by the states, but the federal government provides matching grants for the provision of medical assistance to low-income, aged, blind, and disabled individuals. It has annual expenditures of nearly $14 billion, of which about $286 million are disbursed to motor vehicle accident victims (*1*). Nearly $7 million dollars are saved each year in medical bills for low-income elderly and blind persons because of the 55 mph speed limit. Again, the cost savings attributable to the 55 mph speed limit appear modest in comparison to the total public costs incurred by motor vehicle accidents. However, the injuries averted as a result of the 55 mph speed limit represent only about 2 percent of the 200,000 serious injuries that occur each year, and only a small fraction of those injured are eligible for Medicaid. Further, the majority of all injuries occur on roadways other than those posted at 55 mph (*3*).

Medicare

The Medicare program also has large annual savings attributable to the reduced number of serious injuries because of the 55 mph speed limit. Medicare—as distinct from Medicaid—covers a share of the medical costs for almost all persons aged 65 or above. The Medicare Supplemental Security Income (SSI) program, designed to protect the elderly from high medical costs, pays approximately $140 million annually as a result of highway accidents (*1*). The Supplemental Medical Insurance program obtains about two-thirds of its funding from general tax revenues. The reduced number of minor and serious injuries, some 10 percent of which would have been suffered by elderly persons eligible for the medicare supplemental program, result in annual savings on the order of $5 million.

Savings to Social Service Programs

When lower-income individuals are injured or disabled by an accident, they often turn to public programs to help support them while they are unable to work. These programs also support surviving family members when the household head is killed in an accident. The program costs most directly influenced by motor vehicle accidents include: Supplemental Security Income, Aid to Families with Dependent Children (AFDC), food stamps, and Title XX (which includes a wide variety of social service programs) (*1*). These programs are funded by both the federal and state governments and are administered by state and local governments. The largest impact of reduced injuries and deaths attributable to the 55 mph speed limit occurs in the AFDC program. With approximately 400 fewer deaths and serious, severe, and critical injuries to household heads in the eligible population, this program saves some $1,800,000 annually. The individual impact on other social service programs is small, but together they total approximately $250,000 each year (see estimates in Appendix G).

State governments also fund numerous assistance programs, the costs of which are increased by motor vehicle accidents. A detailed analysis of the cost of motor vehicle accidents in the state of Michigan revealed that motor vehicle accidents increased the program costs of the state's special education, state institutions, and the Crippled Children's Fund by some $1,000,000 annually (*4*). These costs result from the incidence of long-term disabilities such as quadraplegia and severe brain damage. Although comparable estimates for other states are unavailable, extrapolating to the nation on a per capita basis suggests that the costs to all the states for motor vehicle accidents are approximately $28.5 million each year. Assuming that the speed limit reduces these costs in the same proportion in which it reduces Abbreviated Injury Scale (AIS) 4 or 5 injuries, overall costs would be reduced by 4.3 percent, or about $1.2 million each year.

Savings in Public Personnel Costs

Taxpayer costs are directly affected when public employees are killed or injured in automobile accidents. The compensation for serious injuries and sick leave cost for lost time from work for all public employees because of motor vehicle accidents totals about $440 million each year (*1*). Because of the reduced number of injuries attributable to the 55 mph speed limit, these costs are reduced by about $300,000 annually.

Savings in Indirect Public Costs

In addition to the cost of supporting accident victims and their families, many other public costs are indirectly increased because of motor vehicle accidents. These accrue predominantly at the state and local levels. For example, the cost of automobile insurance for state and local governments is determined by the total number of accidents on public roads. Further, accidents on public facilities often result in litigation against local and state governments (Table 18). For example, in 1980, approximately $2 million in judgment costs were assessed against the state of

TABLE 18 Indirect Public Costs Increased by Motor Vehicle Accidents (*1,4*)

Costs	1983 Dollars
Auto insurance	39,500,000
Damaged public facilities	81,700,000
Litigation and judgments	109,000,000
Coroner-medical examiner	10,000,000
Emergency costs (police and fire)	708,000,000
Total	948,200,000

Michigan (*4*). In addition, many accidents result in damage to guardrails, signs, and other public facilities. These costs are usually compensated for by the motorists' insurance, although the state and local governments of Michigan had nearly $2.7 million in uncompensated costs for roadway damage in 1980. The analyses of Michigan's costs by Andary et al. (*4*) have not been replicated by other states; however, extrapolating the estimates in that study on a per capita basis and inflating the costs to 1983 dollars suggest that these costs approach $200 million each year.

Other public costs are also indirectly influenced by the number of accidents, such as the expense of providing emergency services to victims at the scene of an accident and the costs of medical examiners and coroners. The National Highway Traffic Safety Administration estimates that these costs totaled $570 million in 1980, or approximately $700 million in 1983 dollars (*1*). For the most part, however, these costs result from the enormous number of motor vehicle accidents that occur each year, most of which occur on highways posted below 55 mph. In 1981, for example, nearly 11 million vehicles were involved in accidents that included at least property damage, and 3.6 million persons were injured in accidents (*3*). Although the reduced injury accidents attributable to the 55 mph speed limit make up a small fraction of all accidents, severe and fatal accidents increase public costs out of proportion to their actual number.

Two different approaches can be used to estimate these costs. First, assuming that these costs are reduced by their proportion of all injuries and fatalities, the 55 mph speed limit saves about $5 million a year. A second approach is to simply multiply the incidence of injuries and fatalities averted as a result of the speed limit times the average cost of police, fire, medical examiner, and emergency services per accident. By using this approach some $3.9 million in costs are avoided because of the 55 mph speed limit (see Appendix G).

The 55 mph speed limit may indirectly reduce the cost of maintaining the nation's highway system. On irregular pavement surfaces higher speeds may contribute to an increased rate of deterioration. The dynamic load of a vehicle—the static weight plus the downward impact force caused by vertical motions—increases with velocity. When a pavement surface is perfectly smooth dynamic loads are independent of speed. When bumps and cracks appear in a pavement, however, vehicles literally bounce up and down as they travel. This bouncing motion increases the downward impact of the vehicle. Field tests of heavy trucks passing over bumps of less than 1 inch have demonstrated increases of peak dynamic loads of 32 percent of the axle load at speeds of 35 mph. These loads increased marginally at higher speeds (*5*). Although the increased loads were only slightly greater at higher speeds, pavement damage increases exponentially with heavy axle loads, thus multiplying the effect of speed.

Currently, approximately 10 to 12 percent of the federal-aid highway system has irregular surfaces that would directly contribute to a speed-deterioration relationship (*6*). Slower travel on these roads may extend their useful life, although they are already due for resurfacing. On other highways with relatively few bumps and cracks, slower travel speeds may also contribute to extended pavement life. Unfortunately, the relationships between lower speeds, reduced dynamic loads, and extended pavement life have not been established. As a result, estimates of cost savings from lower speeds would require extensive research well beyond the scope of this study.

Revenue Losses

Other estimates of the cost to the federal and state governments associated with motor vehicle accidents include the revenue losses that occur when individuals are killed or lose productive work time because of injuries. The National Highway Traffic Safety Administration reported that some $11 billion is lost to all levels of governments because of the premature deaths of individuals who had been paying taxes (*1*). Although there may be some lost tax revenue when an individual is killed or injured, these estimates ignore the revenue effects of a surplus labor market. Creation of one position will stimulate a chain of moves throughout the labor market as individuals move into new jobs.

Positions rendered vacant for any reason usually do not remain vacant for long, although time is required for advertising and filling the position. These costs are probably smaller than the $11 billion estimate cited previously. For example, assuming that the averted deaths and injuries attributed to the 55 mph speed limit caused job vacancies and that these vacancies remained open for 1 or 2 months each, the revenue losses to all levels of government would be between $3 and 6 million (assuming the average wage and 25 percent of salary in taxes to all levels of government).

PUBLIC COSTS

The 55 mph speed limit also entails costs to the taxpayer. These costs are incurred from the increased enforcement required to maintain compliance and the additional costs associated with installing speed monitoring devices and administering the 55 mph compliance program.

In 1979 the average state spent about $7.3 million to enforce the 55 mph speed limit, or roughly $10.25 million in 1983 dollars (Table 19). This suggests a national cost of about $512 million annually for enforcement of the 55 mph speed limit.

Although these enforcement efforts require taxpayer funding, it is unclear whether they would be significantly reduced by abolishing the 55 mph speed limit. Any speed limit requires some enforcement. In addition, the police have such a diverse and demanding set of responsibilities that elimination of their responsibilities for enforcement of the 55 mph speed limit may only allow them to shift their attention to other traffic safety needs without altering enforcement agency budgets. The breakdown of the cost of enforcement of the 55 mph speed limit clarifies this point. Almost 63 percent of the cost of enforcing the 55 mph speed limit is made up of labor costs and another 14 percent is the cost for vehicle mileage (Table 19). The special costs of enforcement—special enforcement teams, aircraft surveillance, speed monitoring devices, special training, public information, and paperwork—account for only 23 percent of the total. Thus, the avoidable enforcement cost of the 55 mph speed limit is about $118 million dollars each year (23 percent of $512 million).

The revenue generated from traffic tickets may affect total costs considerably. Many states attempt to set the fines such that the revenue generated at least equals the cost of issuing the citation. Generally, these revenues do not actually benefit the police agencies that issue such citations; instead they are funneled into the state's general fund or into special accounts at the local level. Detailed estimates of the revenues generated are not available for all states, but two states were able to

TABLE 19 Average State Cost of 55 mph Speed Limit Enforcement (7)

Category	Cost[a] ($)	Percent
Personnel	4,603,143	62.9
Mileage	1,037,007	14.2
Special teams	704,888	9.6
Aircraft	66,553	0.9
Radar	542,770	7.4
Training	15,819	0.2
Paperwork	183,484	2.5
Public information	73,090	1.0
Other costs	90,173	0.2
Total	7,316,928	100.0

[a]Based on 15-state sample (1979 dollars).

compare the revenues from fines to total enforcement costs (7). New York estimated its total enforcement program for the 55 mph speed limit in 1979 at roughly $30 million and the state earned about $6 million in revenues from fines the same year. Florida spent about $11 million on enforcement of the 55 mph speed limit and earned about $8.4 million in fiscal year 1978 in revenues from fines. Although the average revenue earned per citation is unknown, with about 8 million citations issued in 1983, if the average citation was only $15 the incremental cost of enforcement would be offset by the revenues generated from fines (8 million × $15 = $120 million).

The states must certify to the federal government that they enforce the speed limit and must provide estimates of vehicle speeds on all state highways posted at 55 mph. These administrative costs require at least one full-time equivalent staff position in each state and about one full-time staff position in each Federal Highway Administration regional office. Rough calculations on average salaries and benefits for government employees suggest that approximately $2 million is spent on compliance administration each year. In addition, each state has been required to install speed monitors. These monitors cost about $5,000 each, and because there are about 1,800 of them, the overall cost is about $9 million. Spreading this cost over the 10 years the 55 mph speed limit has been in effect results in an annual cost of slightly less than $1 million per year.

CONCLUSION

In a broad sense, the 55 mph speed limit has many costs and benefits to individual taxpayers that cannot be measured in tax dollars. The risk of accidental death or injury on the highway has been reduced. Fuel conservation lowers the demand for imported oil and improves the nation's balance of payments. Indeed, the principal benefit—reduced pain and suffering—cannot be measured in dollars. The speed limit, however, does impose considerable time costs and these are borne directly by individuals. Because these benefits and costs accrue to all of society much beyond their effects on taxpayer-supported programs, they are discussed in other chapters in this report. In the narrower sense of taxpayer impacts, substantial savings occur in medical and public assistance programs because of the reduced number of fatalities and serious injuries (Table 20). The cost of other programs, notably enforcement, is increased. At the same time, the revenues generated from fines offset some of the additional enforcement costs.

TABLE 20 Total Public Program and Agency Budget Savings and Costs of the 55 mph Speed Limit (1983 dollars)

Program/Budget Savings and Costs	Millions ($)
Savings	
Medical and family assistance programs	52.0
Social service programs	3.5
Public personnel costs	0.4
Indirect public costs	3.9 to 5.0
Tax revenues	3.0 to 6.0
Total	62.8 to 66.9
Costs	
Enforcement	118.0
Compliance administration	2.9
Total	120.9

REFERENCES

1. *The Economic Cost to Society of Motor Vehicle Accidents*. Appendix A. NHTSA, U.S. Department of Transportation, Jan. 1983.
2. Budget of the United States, Fiscal Year 1983. Executive Office of the President, Office of Management and Budget, pp. 5-143.
3. *National Accident Sampling System Report*. NHTSA, U.S. Department of Transportation, 1982.
4. J.J. Andary, J.D. Flora, D.F. Huelke, and J. O'Day. *Estimates of Direct Costs to the State Resulting from Traffic Accidents*. Highway Safety Research Institute, University of Michigan, Ann Arbor, Aug. 1981.
5. A.P. Whittemore, J.R. Wiler, P.C. Shultz, and D.E. Pollock. Dynamic Loads of Heavy Highway Vehicles. *NCHRP Report 105*. TRB, National Research Council, Washington, D.C., 1970.
6. *The Status of the Nation's Highways: Conditions and Performance*. FHWA, U.S. Department of Transportation, 1983.
7. R. Sostkowski. *Determining the Cost of Enforcing the 55 mph Speed Limit*. Report DTNH22-81-c-05010. NHTSA, U.S. Department of Transportation, April 1981.

6
Energy Savings

In 1973 the United States, with 6 percent of the world's population, consumed 35 percent of all the energy produced. Although the energy shortages caused by the 1973 Arab oil embargo took most Americans by surprise, many industry and government officials had been aware of the nation's growing energy problems. Between 1960 and 1973 the demand for petroleum increased by 76 percent while domestic production increased by only 15 percent. Domestic production of petroleum had peaked by 1973 and imported oil was meeting an ever larger share of the nation's fuel needs as shown in Figure 20 (*1*,*2*).

Moreover, spot shortages of gasoline had begun to occur several months before the Arab oil embargo in October 1973. Analysts had predicted a possible shortage of heating oil for the winter of 1972-1973, and in response, refineries had increased the output of distillates at the expense of gasoline. The winter turned out to be exceptionally mild and this production imbalance, combined with a rapid increase in the demand for gasoline in the summer of 1973, produced spot shortages in some parts of the country. The demand for all petroleum products for the winter of 1973-1974 was projected to be up by 1.2 million barrels per day (BPD), whereas domestic production was projected to rise by only 200,000 BPD. Imports were expected to ensure that supplies remained above the critical level. The embargo, therefore, occurred at a particularly inopportune time. Demand was estimated to be 19.1 million barrels

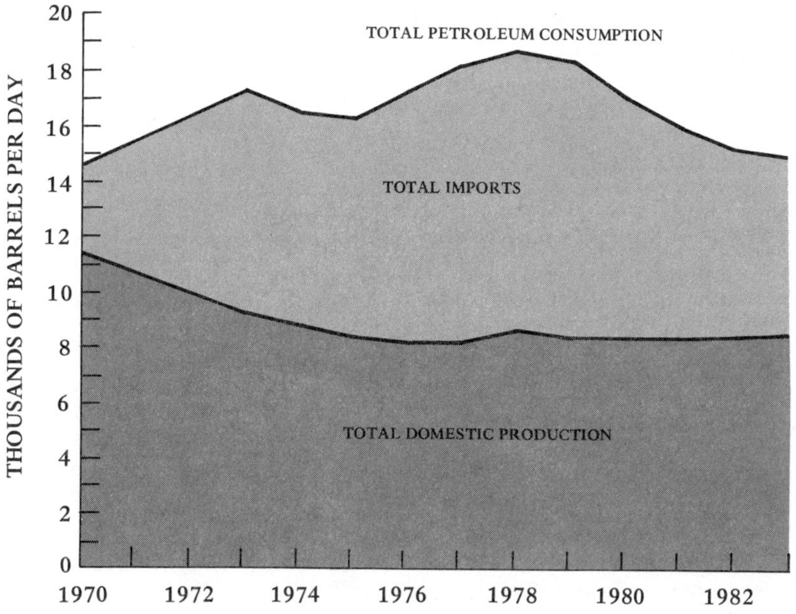

FIGURE 20 Trends in petroleum imports, 1970–1983 (*1*, *2*).

per day, but supply was only 17.3 million barrels per day—a 10 percent shortfall (*3*).

The lines at gasoline stations were the most visible manifestation of the energy crisis of 1973-1974, but heating oil supplies, agricultural activities, and industrial production were also threatened by the crisis. Because substantial amounts of travel are noncritical and because considerable excess capacity exists in the transportation network, discussions of energy conservation policy often focused on the transport sector. Transportation also held promise for major fuel savings, because it accounts for approximately one-half of total petroleum consumption. Motorists were urged to reduce unnecessary travel, combine trips when possible, develop ridesharing arrangements, and drive at slower speeds. Many motorists voluntarily reduced their speeds to conserve the fuel that was available. In addition, many states took unilateral action to reduce speed limits (*4*).

In January 1974 Congress passed the Emergency Highway Energy Conservation Act, which set the maximum speed for all vehicles at 55 mph. States that refused to lower their speed limits to 55 mph within 90 days risked losing federal highway aid. By March 1974 all states had complied. The legislation was designed as a temporary measure and was to expire in April 1975 or whenever the President determined that a fuel crisis no longer existed. Although the embargo was lifted in April 1974, the President did not lift the speed restriction.

EXPECTED SAVINGS

Before the 55 mph speed limit was enacted several studies had been conducted that documented the relationship between speed and fuel consumption. The U.S. Department of Transportation (DOT) (*5*) found that restricting all vehicles to 55 mph would save nearly 200,000 BPD of oil, which represents slightly more than 1 percent of total petroleum demand. Maximum savings of approximately 500,000 BPD could be achieved if speed limits were reduced still further to 50 mph, but the time costs associated with that speed were considered unacceptable (*5*).

A theoretical computation of the effect of the 55 mph speed limit on energy conservation was undertaken in a National Science Foundation study (*3*). Fuel consumption was analyzed for three categories of corporate motor vehicles, taking into account the distribution of speeds on main rural roads and total miles traveled for each vehicle type. Assuming that all vehicles complied with the 55 mph speed limit, it was concluded that about 200,000 BPD of fuel would be saved. However, if the speed limit was set at 60 mph (or if all vehicles traveled 5 miles above the 55 mph speed limit) savings would be only one-half as large (*3*).

Gasoline consumption declined 3.7 percent in 1974, or by about 255,000 BPD. Because consumption had been increasing by 4 to 6 percent annually, this decline was especially significant. However, part of the decline could be traced to reduced travel, and part could be traced to slower driving—not in response to the 55 mph speed limit, but in response to higher fuel prices and gasoline shortages.

CURRENT FUEL SAVINGS

The energy crisis increased consumer demand for more fuel efficient automobiles, and the fuel efficiency of the nation's automobile fleet has improved by at least 24 percent since 1973. Scaled down engines, improved gear ratios more suitable for 55 mph, smaller and lighter vehicles, and radial tires have contributed to improved fuel economy. Downsizing of the American automobile fleet has been a more important energy conservation measure than the 55 mph speed limit. However, the speed restriction negated some of the appeal of powerful, large cars and might have contributed to consumer acceptability of small cars with smaller engines.

In a recent comprehensive evaluation of the energy conservation benefits of the 55 mph speed limit, Mason and Zub (*6*) assembled fuel performance information on 593 vehicles. Fuel performance data on passenger cars were obtained from independent testing organizations, manufacturers, and government agencies. Fuel economy data on commercial vehicles were derived from earlier FHWA studies and from test results from truck drivers who attempted to show that they could get better fuel economy at higher speeds (*7,8*).

As revealed in previous studies, fuel consumption increases rapidly at higher speeds. Fuel consumption is 14 to 31 percent greater for passenger cars traveling at 70 mph compared with those traveling 55 mph. Reducing speeds further to 45 mph would save only 5 to 11 percent more fuel. Moreover, the data also indicated that newer cars (1980 models) averaged 10.6 percent better fuel economy than vehicles in the same weight classes built before 1976.

To measure the aggregate energy savings of the 55 mph speed limit, data were collected on the annual mileage traveled by vehicles of different vintages, the distribution of vehicle speeds on the roads in 1972 and in 1979, and the fuel consumption rates for each vehicle at different speeds (*6*). As vehicles grow older, they are driven fewer miles, and as they are retired they become a smaller portion of the fleet. Given survivability and registration data, the proportion of total vehicle miles accounted for by each model year was estimated (see Table A-28 given in Appendix I). By comparing fuel consumption for the 1979 fleet of passenger cars at the speed limits of 1979 and 1972, Mason and Zub estimated fuel savings of 112,000 BPD (Table 21). If 90 percent compliance with the 55 mph speed limit could be achieved, the savings would be 232,000 BPD of fuel. Similar analyses for light trucks resulted in an estimate of 22,000 BPD of fuel saved with potential savings of 48,000 BPD at 90 percent compliance.

Heavy truck and bus fuel economy was analyzed separately. Constant-

TABLE 21 Current Fuel Savings as a Result of the 55 mph Speed Limit (BPD)

	1979	1983[a]
Passenger cars	112,000	110,000
Light trucks	22,000	26,000
Heavy trucks and buses	26,000	31,000
Total	160,000	167,000

[a]Updated estimates based on Mason and Zub (*6*).

speed fuel economy estimates for commercial trucks were developed by performing regression analysis on data obtained in the Double Nickel Challenge tests. The regression equation was:

$$\text{Fuel Consumption Rate} = C_1 + C_2 w + C_3 v^2$$

where w is truck weight and v is selected speed (mph). This form of the model takes into account rolling resistance (C_2w) and aerodynamic drag (C_3v^2) in determining fuel economy.

Bus fuel economy figures were based on earlier research. Together, buses and heavy trucks were found to be saving 26,000 BPD of fuel due to slower highway travel speeds.

In the 2 years since that study (*6*) was released, there have been additional changes in the vehicle fleet and the volume of travel that have altered the impact of the 55 mph speed limit on fuel savings. It was not possible to completely reestimate the savings. The original work relied on easily accessible secondary data sources that are no longer readily available. Nevertheless, it is not difficult to derive a reasonable estimate of the current impact of the 55 mph speed limit on fuel consumption.

Total travel has increased, although modestly, since 1980. Traffic volumes increased by about 2 percent annually during the past 3 years. Thus, based on the original estimate, travel on roads posted at 55 mph was approximately 750 billion vehicle miles in 1983. The share of travel accounted for by passenger cars has declined in the past 3 years from 75.9 percent to 71.5 percent on roads posted at 55 mph. To recalculate fuel savings, data are needed for the age distribution of the vehicle fleet and the mean fuel economy rates of the more recent models.

Although it is not feasible to collect the kind of detailed fuel performance data amassed for the original study, analysis of fuel economy performance suggests a linear improvement over time. A simple extrapolation of the trend yielded fuel economy performances for 1980-1983 vehicles under the two speed distributions. The newest models (1983) are estimated to be about 30 percent more fuel efficient than 1979 cars, the last year for which data were reported in the original study. Combined highway and city automobile fuel economy reported by the manufacturers has improved approximately 35 percent during this period, so this extrapolation is probably not far from the mark.

For simplicity it has been assumed that the distribution of annual mileage by age of car has not changed since 1980 (estimates shown in Table A-29, Appendix I). Because older cars are being kept longer and new car sales have slumped in recent years, the overall fuel economy is probably somewhat overstated. However, average speeds have also gradually increased and this translates into poorer fuel economy for the current passenger car fleet.

If the 1983 fleet of passenger cars traveled at the speeds that prevailed before the 55 mph speed limit was enacted, then 1,670,000 barrels of fuel would have been consumed daily on highways posted at 55 mph. Traveling at slower speeds reduced consumption to 1,560,000 BPD, or a savings of 110,000 BPD. This is slightly less than the 112,000 BPD savings projected for the 1979 passenger car fleet (Table 21). The reason for the reduction is simply that fuel economy improvements have risen faster than travel.

Commercial and light trucks have increased their share of travel on

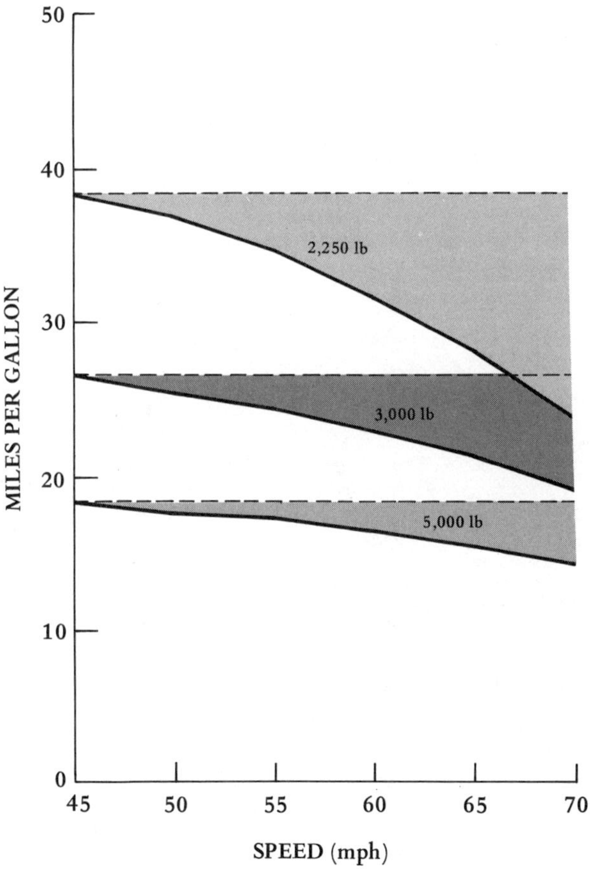

FIGURE 21 Declining fuel economy with increased speed by weight of vehicle (*6*).

highways posted at 55 mph during the past few years. From 24.1 percent in 1980, their share increased to 29.5 percent in 1982. If the energy conservation benefit from trucks traveling 55 mph remained relatively unchanged, then adjusting for the increase in truck traffic yields an estimate of 26,000 BPD saved from light trucks traveling 55 mph and 31,000 BPD saved by heavy commercial vehicles. The total savings for all vehicles is 167,000 BPD, or about the same as estimated for the 1979 fleet.

The amount of motor fuel saved as a result of the 55 mph speed limit will increase over time as the vehicle fleet is made up increasingly of small (downsized) vehicles. At higher speeds fuel-consumption rates rise more rapidly for small cars than for large cars (Figure 21).

In recent years the national concern about fuel savings has diminished as petroleum stocks have increased following a period of decreasing demand by industrial nations and increased supplies from some nations producing oil and seeking additional oil revenues. U.S. dependency on

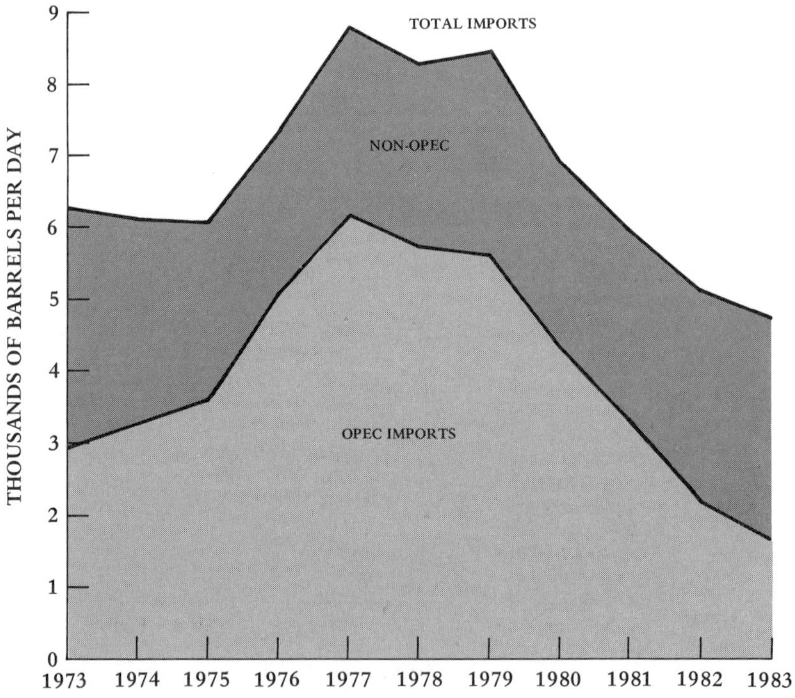

FIGURE 22 Trends in petroleum imports from OPEC nations, 1973–1983 (*2*).

Arab Organization of Petroleum Exporting Countries' (OPEC) sources actually rose substantially after 1973, but in the past few years it has fallen substantially (Figure 22).

However, should another energy crisis occur, the energy savings attributable to the 55 mph speed limit could be important. Even though the current fuel savings are only about 1 percent of total consumption, the energy crisis of 1973 resulted from a shortfall of only 10 percent of demand. Moreover, the fuel savings estimated here have a current market price of about $2 billion annually. This represents a substantial reduction in the balance of payments deficit.

REFERENCES

1. H.R. Millie, ed. *Minerals and Materials: A Monthly Survey*. Bureau of Mines, U.S. Department of the Interior, 1976, p. 7.
2. *Monthly Energy Review*. Energy Information Administration, U.S. Department of Energy, Nov. 1983, pp. 32-33.
3. H.B. Schecter, et al. *Policy Assessment of the 55 mph Speed Limit*. The MITRE Corporation, McLean, Va., 1975.
4. *55 Fact Book*. NHTSA, U.S. Department of Transportation, 1977.
5. A. French and H. Bishop. *Analysis of Fuel Savings Through Reduced Speed Limits*. HHP-40. FHWA, U.S. Department of Transportation, 1973.
6. R.L. Mason and R.W. Zub. *Highway Fuel Economy Study*. DOT-TSC-NHTSA-81-18. Transportation Systems Center, NHTSA, U.S. Department of Transportation, 1981.
7. *The Double Nickel Challenge to the Fuel Pump: Voluntary Truck and Bus Fuel Economy Program NRD-20*. NHTSA, U.S. Department of Transportation, 1979.
8. W.L. Page, A. French, and J.E. Ullman, "Estimated Highway Fuel Savings in 1975." Presented at Governor's Highway Safety Conference, Dover, Del., Oct. 1976.

7
Travel Time and Cost Impacts

The slower and more uniform speeds that resulted from the 55 mph speed limit have saved lives and fuel, and have reduced the number of injuries, the nation's medical bill, and the costs of government programs. Americans have paid a price for these benefits. Motorists now spend about one billion extra hours on the highways each year because of slower driving speeds. Those who use the highways to earn a living, especially truckers, are less productive because some hauls now take longer to complete. Although the added time is minimal in some cases and may go unnoticed by drivers, this is not true in some circumstances. For example, in rural areas where drivers often travel long distances on uncongested, relatively safe, well-designed highways, the 55 mph speed limit creates frustration and resentment.

The effect the speed limit has had on certain highways and regions of the country is examined in this chapter. After reviewing the time costs of slower driving, the potential for selective, moderate relaxation of the speed limit on the safest portions of the nation's highway system is discussed.

LEVEL AND TYPE OF TIME COST IMPACTS

Roughly 577,000 miles of highway are currently posted at 55 mph. Although information on how these highways were posted before the enactment of the 55 mph speed limit is not available, speed limit reductions were greatest on the Interstates, where daytime speeds of 75 mph had been permitted in some states (*1*). Speed limits were lowered on many arterial and collector roads as well, but these reductions were usually smaller. Also, the design standards and other characteristics of lesser roads effectively limited how fast most drivers could travel.

Today, more than 700 billion vehicle miles of travel (VMT) takes place annually on highways posted at 55 mph. Almost 82 percent of this traffic consists of automobiles or light trucks being used for noncommercial purposes. Commercial truck traffic accounts for nearly all the remainder, while buses represent less than 0.4 percent of VMT on highways posted at 55 mph (Table 22).

Added Time for Personal Travel

The most widely felt cost of the 55 mph speed limit is the additional time motorists spend on the road. The time spent driving may be taken away from other, more productive or more pleasant, activities. Most people consider a trip as an obstacle to getting to where they want to be. The time people spend traveling is, therefore, viewed as part of the price that must be paid to do something else or to be somewhere else.

In assessing the importance of the additional time expended because of the 55 mph speed limit, it is useful to separate travel related to earning

TABLE 22 Estimated Vehicle Miles of Travel Affected by the 55 mph Speed Limit

	1982 VMT 55 mph Posted Roads (millions)			
Highway System	All Vehicles	Passenger Cars and Light Trucks for Personal Use	Trucks and Light Trucks for Commercial Use	Buses
Rural Interstate	140,778	105,913	34,370	495
Rural arterial	223,903	181,371	41,632	900
Rural collector	106,950	88,056	18,325	569
Urban Interstate	157,974	134,716	22,790	468
Urban freeway and expressways	73,999	64,449	9,239	311
Total	703,604	574,505	126,356	2,743

NOTE: Estimates of total travel on highways posted at 55 mph provided by the Federal Highway Administration, Highway Performance Monitoring System. Vehicle miles of travel on urban arterials posted at 55 mph assumed to have been unaffected by the 55 mph speed limit.

a living from travel for recreational or social purposes. The importance of saving time differs depending on the type of trip being made.

Work-Related Trips

Added time on the road results in some readily identifiable, economic costs to individuals and businesses. Time spent in work-related travel is presumed to be worth the value of what could have been produced if the travel were not necessary. This can be approximated by the wage rates of the driver and other vehicle occupants.

Unpaid time spent commuting to or from work also is usually assumed to have positive economic value, but worth something less than the wage rate as it entails no direct loss of productive time. More than 80 percent of work-related travel is spent commuting to and from jobs. The consensus from the many studies that estimated the value of travel time indicates that commuting time is worth one-third to two-thirds of the average hourly wage (*2*). Work-related trips (including commuting and driving for work) account for about 38 percent of all personal vehicle travel (Table 23).

Although work-related driving makes up a considerable share of total personal vehicle travel, the majority of these trips are quite short. The average trip length for commuters is about 10 miles each way (*3*). Trips this short are unlikely to take place entirely on highways posted at 55 mph; however, even if such trips were to take place on these highways the additional travel time due to slower speeds would amount to slightly less than 1 minute (comparing 1973 and 1983 average rural Interstate speeds).

TABLE 23 Percent of Vehicle Trips and Travel for Passenger Vehicles (*3*)

Trip Purpose	Trips (%)	Travel (%)
Earning a living		
Home to work	27.8	30.4
Work related	5.1	7.3
Subtotal	32.9	37.7
Family and personal	32.3	22.9
Civic, educational, religious	6.4	4.7
Social and recreational	19.3	24.0
Other	9.1	10.7
Total	100.0	100.0

Studies of passenger tolerance for delays during commuting suggest that most commuters ignore time losses of 2 to 4 minutes, although they probably take seriously additional time losses (*4*). This indicates that for many commuters slowed down by the 55 mph speed limit, the time losses would be largely discounted. On the basis of average trip lengths and average speeds, most personal vehicle travel slowed down by the 55 mph speed limit involves time losses of less than 3 minutes (Table 24). The time losses per trip in Table 24 would vary by highway class and region depending on the posted speed before enactment of the 55 mph speed limit. Nonetheless, on average, the time losses per trip tend to be fairly small.

Nonwork-Related Trips

The preponderance of all passenger vehicle travel (62 percent) is for personal business, shopping, and recreational purposes. The value of time spent making these types of trips is difficult to estimate. Extra time spent making recreational trips is probably worth less than the added time commuting between home and work. Most people would pay something to reduce their travel time for nonwork trips, but estimates of the economic value of this type of travel time remain uncertain. It seems reasonable to conclude that drivers making nonwork trips are most concerned by the speed limit when making lengthy, routine trips and occasional long-distance pleasure trips where the extra time on the road can be substantial. The cost of the 55 mph speed limit is probably demonstrated by motorist frustration on automobile trips involving more than 1 or 2 hours of driving. As noted in Table 24, these trips account for 34 percent of total personal travel.

TABLE 24 Estimated Passenger Time Cost by Length of Trip

Length/Miles	Additional Minutes of Driving	Passenger Travel (%)
11–20	0.7 to 1.25	32
21–30	1.26 to 1.90	17
31–50	2.00 to 3.10	17
51–100	13.20 to 6.20	15
More than 100	More than 6.20	19

NOTE: See Table A-31, Appendix I, for calculations.

Commercial Time Impacts

Trucking

In 1982 truckers drove 126.4 billion miles on highways posted at 55 mph; approximately 60 percent of this travel was on Interstates (Table 22). The 55 mph speed limit lowers the productive capacity of the nation's trucking industry and raises labor costs. Part of the increased operating expenses is passed on to consumers in the form of higher prices for goods transported by truck, and part is absorbed by those who own or drive the nation's trucks. Nevertheless, there are some offsetting benefits from driving slower, especially in lower truck maintenance costs and fuel savings.

The trucking industry is divided on the issue of the net effects of the 55 mph speed limit. The large common carriers and their primary trade organization, the American Trucking Association, publicly support the speed limit for both safety and economic reasons. In contrast, the independent owner-operators oppose the 55 mph speed limit. The larger carriers and private truck fleets may be inclined to take a more purely business approach to the need to conform equipment and schedules for compliance, and, in so doing, realize some of the offsetting benefits of slower speeds, such as fuel conservation and lower maintenance. The owner-operator may or may not reach the same conclusion regarding the business decision, but because he is the driver as well as the owner, personal feelings of frustration at being "held down" may play a part. Also, in the case of the owner-operator, loss of vehicle productivity is the same as a loss in personal productivity.

Safety regulations govern the number of hours truck drivers can travel each day, and slower speeds therefore mean that fewer miles can be driven per legal workday. An analysis provided to this study by the Independent Drivers Association indicates that a typical owner-operator earns about $1 per mile and that an owner-operator driving 65 mph can complete a 435-mile trip in 7 hours and 43 minutes; whereas if he obeyed the speed limit, the same trip would take 8 hours and 48 minutes (according to U.S. Department of Transportation estimates that include meal and other rest stops). By driving faster, the independent trucker can gain an hour or $65. The 55 mph speed limit, according to the Association's calculations, causes some owner-operators to lose as much as $15,000 annually. The effect on common carriers is to require more drivers and equipment to accomplish the same number of miles.

Although the 55 mph speed limit costs time, it saves fuel. A test comparing the performance of two long-haul tractor trailers demonstrated that reducing the speed from 60 mph to 55 mph saved 10 percent of fuel consumption on hilly terrain. Reducing the speed to 50 mph, however, saved only 3 percent more fuel (*5*). A United Parcel Service (UPS) test in 1977 suggested even more dramatic savings. A combination truck traveling 55 mph used 32 percent less fuel than a similar truck traveling 65 mph. If speeds were 50 mph instead of 55 mph, fuel consumption was still less, and UPS concluded that a reduction of only 5 mph could save 1,500 to 2,000 gallons of fuel per year for every truck traveling 100,000 miles (*6*).

In addition to fuel costs, the 55 mph speed limit reduces truck maintenance costs. A staff engineer for the Regular Common Carrier Conference reported the following potential cost savings (*7*):

- Lower brake line pressure for stopping reduces lining and drum wear, with less scoring and heat checking, and reduces the effect of brake imbalance between tractor and trailer, a major cause of jackknifing;
- Tire wear is reduced 16 percent from operations at 65 mph;
- Engine life is extended 12 to 20 percent depending on the rpm reduction;
- Engine overhaul and rebuilding cost savings amount to approximately $600 and $1,000, respectively;
- Use of lighter transmissions, drive axles, and tires needed by a smaller engine; and
- Regular annual maintenance reductions (other than those listed above) total between $800 and $900.

Some carriers also report reduced insurance rates from speed control programs (*8*). Further, there is some evidence that the safety of trucking operations has been improved by the 55 mph speed limit. A study of truck accident data in four states revealed a 15 percent decline in truck and car crashes despite a 13 percent increase in truck travel (*9*). A study of toll-road data revealed a 16.6 percent decline in truck accidents in 1974 while traffic declined only 1.2 percent in 1974 (*10*).

Whatever net economic effects driving slower has on time, fuel, and maintenance costs, the actual change in driving speeds for trucks as a result of the 55 mph speed limit is probably smaller than that for cars. In many states trucks were restricted to slower maximum speeds than passenger cars before 1974, and, therefore, the change in the speed limit did not affect the speed trucks could legally travel. Before 1974 truck speeds on main rural Interstates were almost 5 mph less than the average speed for passenger cars (*11*). But by 1978 trucks and passenger cars traveled at roughly the same average speed (*12*). Average truck speeds declined only 2 mph compared to a 7 mph reduction for automobiles. However, given the much larger amount of miles driven by commercial vehicles, this 2 mph reduction amounts to a greater loss of productivity.

An ongoing survey of truck drivers corroborates this finding. In this survey drivers at truck stops are questioned about their last five trip segments. Owner-operators report averaging 63 mph and other truck drivers report that they average 62 mph on the open road. Since this survey began in 1978, more than 88,000 drivers have been interviewed, and there has been little change in driver-reported speeds during the period (*13*). Although this survey provides no data on pre-1974 speeds, it is clear that the speeds reported since 1978 would be in violation of the speed limit and subject the driver to the risk of citation, fine, and, if continued, possible loss of license; whereas the same speeds were often legal in 1974.

In 1975 federal legislation was passed that allowed states to increase maximum weight limits on the Interstates to partly compensate truckers for the productivity loss stemming from the 55 mph speed limit. As a result about 12 states increased their weight allowance. Other states chose not to enact the higher weight difference, and some already had grandfather clauses that permitted heavier trucks. It has been argued that, because of this compensating policy, truckers are more productive in 1983 than in 1973.

Finally, the impact of the 55 mph speed limit on trucking is greatest on the Interstates and less to nonexistent on other highway systems. By 1975 both single-unit and combination trucks were again traveling as fast on rural primary and secondary roads as they were before the 55 mph speed limit was imposed (*14*).

If only time and fuel consumption rates are taken into account, then operating costs in 1975 were 2 to 3 percent higher for Interstate truck traffic than in 1973, due to the 55 mph speed limit. If offsetting maintenance, safety, and other savings are included, the net increase in cost is even less (*14*).

The trucking industry has been highly volatile in the past few years, and many firms have failed. Deregulation and the recent recession have contributed to the economic turmoil facing the nation's trucking industry. Although the 55 mph speed limit is often attacked by some segments of the trucking industry, the net effect of this policy appears neglible compared to that of the other economic changes that have recently confronted this industry.

Buses

Buses travel about 2.7 billion vehicle miles each year on highways posted at 55 mph. As with truck and automobile traffic, buses slowed down most on rural Interstates and less on other systems. Bus speeds declined slightly less than automobile speeds after the 55 mph speed limit was introduced.

Bus riders spend an extra 46 million person-hours each year because of slower bus speeds (Table 25). Many of these additional hours are spent on recreational trips, or trips to visit family or friends. Few intercity bus riders travel on business and most commuter bus traffic occurs at times or in places where speeds above 55 mph would be difficult to achieve in any case.

Some bus companies report that labor productivity has declined, as average speeds are now about 10 percent below pre-55 mph speed limit levels. The 55 mph speed limit required intercity bus operators to lengthen their schedules, reassign drivers, and in some cases, hire additional personnel. Greyhound, which operates about 60 percent of the regular-route intercity services, estimates that labor productivity declined 10 percent because of the 55 mph speed limit. If Greyhound's experience is typical, then the intercity bus industry incurs approximately $50 million annually in added labor costs. Partially offsetting these increased operating costs are savings on fuel and maintenance expenses (*15*).

Summary of National Time Cost Impacts

Although those who drive on the nation's highways must spend more time making their trips because of the 55 mph speed limit, these time costs must be weighed alongside the sizable safety benefits discussed in

TABLE 25 Additional Passenger Hours of Travel (1982)

Vehicle Type	Passenger Hours (thousands)	Percent of Total
Cars and light trucks	807,526	77.8
Commercial trucks	184,068	17.7
Buses	45,966	4.4
Total	1,037,560	100.0

NOTE: See Tables A-33 and A-34, Appendix I, for calculations.

Chapters 3 and 4 when evaluating the net desirability of the 55 mph speed limit. Comparisons of average speeds between 1973 and 1982 are inexact for all the highway systems affected by the 55 mph speed limit; nonetheless, it appears that travel in 1982 on highways posted at 55 mph required about 1 billion additional hours of passenger time when compared to the speeds of 1973 (Table 25). The majority of this additional travel time (78 percent) is required of passengers in personal vehicles.

The total additional travel time requires all users of the highways to spend more time traveling each year. On a per-licensed-driver basis, as an approximation of adult users, motorists spend an additional 7 hours driving per year. On a per-person basis (1983 population) Americans spend 4.5 hours more traveling on highways posted at 55 mph each year.

The additional travel time requires about 350,000 hours of additional driving for each life saved (plus avoiding an additional serious, severe, or critical injury). This works out to roughly 40 years of additional driving per life saved and serious injury averted. Although 40 years seems like a tremendous amount of driving, the average life expectancy of motor vehicle accident victims in 1982 was about 41 years (*16,17*).

Although no meaningful comparison can be made between the value of a year of life and the value of a year of driving time, the similarity between the number of years of extra driving time and the number of years of life saved, nevertheless, provides a useful vantage point for assessing the central trade-off involved in the 55 mph speed limit. Although the number of years involved on both sides of the issue are similar, most people would probably agree that an extra year of life is worth the societal cost of an extra year spent driving. This comparison thus suggests that the time-life trade-off imposed by the 55 speed mph limit is one that many would find worthwhile. In addition, this comparison does not include the reduced time spent recovering from serious injuries or being incapacitated.

However, the time and safety trade-off varies among the different road systems and users. On the safest highways, the Interstates, the trade-off is relatively more costly than on other, less safe roads. There are also regional differences in the distribution of time costs, which help explain regional differences in support of the 55 mph speed limit.

Geographic Distribution of Travel Time Costs

The additional travel time imposed by the 55 mph speed limit is unevenly distributed. Residents in more sparsely populated states spend more time driving over highways posted at 55 mph, and their trips are typically longer because activities are located at greater distances. As a result, drivers in rural areas may bear a greater share of the time cost burden than drivers elsewhere.

Highway travel statistics reflect this uneven burden (*18*). For example, in many Western states more that 15 percent of all highway travel takes place on rural Interstates. By contrast, some states in the Northeast average about 6 percent of their traffic on rural Interstates (Figure 23).

Although the travel time burden is not equally shared, neither are the attendent safety benefits. Highway fatalities declined more sharply in 1974 in those same regions where the time costs are greatest. If the safety benefit today is similarly distributed, then there is a rough parity in the regional distribution of time costs and safety benefits (Figure 24).

Time Cost and Safety Benefits on Rural Interstate Highways

Rural Interstates are particularly important for personal travel and commerce because they connect distant cities. These highways had the highest speeds in 1974; thus when speed limits were lowered the time losses were greatest on rural Interstates. Because these highways also

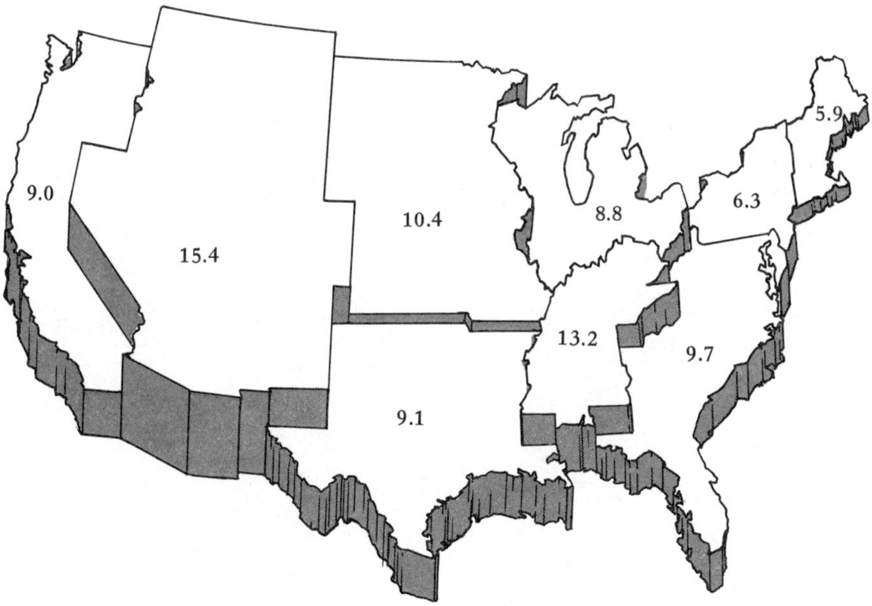

FIGURE 23 Travel on rural Interstates as a percentage of total travel (*18*).

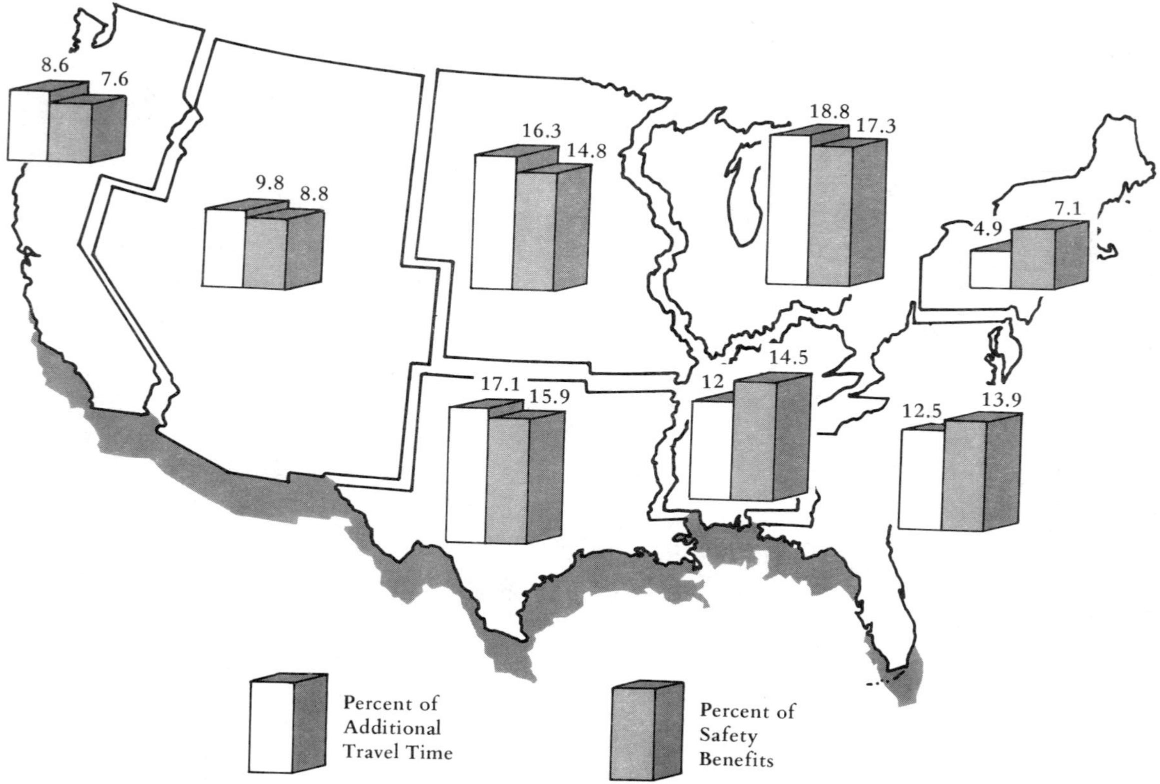

FIGURE 24 Regional distribution of time costs and lives saved on main rural highways.

carry the majority of long-distance travel, the frustration with slower travel speeds is felt most directly by users of this system. Rural Interstates also have a relatively good safety record. The fatality rate of 1.53 deaths per 100 million miles in 1982 on rural Interstates was lower than all other systems except for the urban Interstates (*19*).

When the lives saved by the 55 mph speed limit and the additional driving time required are distributed across road systems, comparisons can be made between different road systems by using available speed and accident data. On this basis the speed limit currently saves about 525 lives each year on rural Interstates (Table 26). Similar calculations for other highway systems principally affected by the 55 mph speed limit indicate that the limit is much more costly in terms of the extra driving time for each life saved on rural Interstates than for the other road systems. Motorists on rural Interstates spend about four times as much added travel time for each life saved as do motorists on the other highway systems (in aggregate), as indicated by the data in Table 26.

SUMMARY

The 55 mph speed limit imposes considerable time costs on the nation's drivers. Motorists spend about 1 billion additional hours traveling each year because of this speed limit. Truck traffic has been particularly affected because many trucks travel great distances on high-speed intercity routes. Nevertheless, the net effect on trucking costs may not be large. Truck speeds were reduced much less than automobile speeds, and in some states truckers benefited from eased restrictions on sizes and weights. Further, some operating costs for trucks are reduced when driving at lower speeds.

For the average motorist, travel time costs are greater in the West. Westerners do more driving on Interstate routes than do Easterners.

TABLE 26 Additional Travel Time Per Life Saved for Highway Systems Posted at 55 mph

Highway System	Estimated Fatality Reductions	Additional Passenger Travel (years)	Years of Driving per Life Saved
Rural Interstate	525	50,800	97
Urban Interstate and freeways	375	11,600	31
Rural arterials	1,650	46,800	28
Rural collectors	650	9,200	14

Westerners travel more miles on highways posted at 55 mph and encounter more open roads formerly posted at speed limits 15 and 20 mph higher. Westerners also benefit from improved safety; however, because fatal accidents have low probabilities, these benefits may not be recognized by individual motorists.

Motorists using Interstate routes, which are among the nation's safest roads, expend more additional travel time for each fatality averted by the 55 mph speed limit than do motorists using other roads. On the rural routes of the Interstate system, the 55 mph speed limit requires about four times as much additional travel time per life saved than it does on other highways posted at 55 mph. This trade-off may underpin some of the objections being voiced concerning the 55 mph speed limit.

REFERENCES

1. Highway Users Federation. *State Laws Governing Motor Vehicles Speed Limits*. Washington, D.C., 1973.
2. N.G. Yucel. "A Survey of the Theories and Empirical Investigation of the Value of Time Savings." Staff Working Paper 199. International Bank for Reconstruction and Development, Washington, D.C., 1975.
3. "Purposes of Vehicle Trips and Travel." Table 1. *1977 Nationwide Personal Transportation Study*. FHWA, U.S. Department of Transportation, 1980.
4. A. A. Altshuler. *The Urban Transportation System*. MIT Press, Cambridge, Mass., 1979.
5. A. J. Broderick. *Fuel Consumption of Tractor-Trailer Trucks as Affected by Speed Limit and Payload Weight*. U.S. Department of Transportation, 1975.
6. United Parcel Service. *Fuel Summary Speed Test Report (55 mph vs. 65 mph) to the Maintenance Committee of the Regular Common Carrier Conference*. Columbus, Ohio, 1977.
7. "Double Nickle Value Greater than Ten Cents." *Ohio Truck Times*, Vol. 28, No. 2, Spring 1979.
8. J. F. Weiss et al. "An Analysis of Commercial Vehicle Speed Control Concepts." *SAE Technical Paper Series*. Nov. 1982.
9. L. A. Zaremba and M. J. Ginzburg. "The 55 mph Limit and Front to Rear Collisions Involving Semis and Large Trucks." *Accident Analysis and Prevention*, Dec. 1977.
10. K. Campbell, R. Scott, and S. Tolkin. *Highway Safety Effects of the Energy Costs on U.S. Toll Roads*. U.S. Department of Transportation, 1976.
11. *Highway Statistics: 1973*. Table VS-1. FHWA, U.S. Department of Transportation, 1974.
12. "Truck Speeds Equal Car Speeds on Rural Interstates." *Fuel Energy News*. U.S. Department of Energy, 1979.
13. Association of American Railroads. *A Statistical Overview of the Intercity Trucking Industry 1981*. Washington, D.C., 1982.
14. D. B. Kamerud. "The 55 mph Speed Limit: Costs, Benefits and Implied Trade-offs." *Transportation Research*, Vol. 17A, No. 1, 1983.

15. Transportation Systems Center. *Effect of Variation of Speed Levels on Intercity Bus Fuel Consumption, Coach and Driver Utilization and Corporate Profitability*. Cambridge, Mass., Nov. 1975.
16. *Statistical Abstract of the United States, 1982-3*. Table 107. Expectation of Life and Expected Deaths. Bureau of the Census, U.S. Department of Commerce, 1979.
17. *Fatal Accident Reporting System*. NHTSA, U.S. Department of Transportation, 1983.
18. *Highway Statistics: 1982*. Table UM-2. FHWA, U.S. Department of Transportation, 1983.
19. *Highway Safety Performance 1982 - Fatal and Injury Accident Rates on Public Roads in the United States*. FHWA, U.S. Department of Transportation, 1983.

8
Public Opinion

Enforcement of traffic laws, as with any law, requires that most people support the basic intent of the rule. The majority of Americans, when asked whether they support or oppose the 55 mph speed limit, endorse retention of the law. Gallup polls in 1974 and 1982 indicate virtually the same level of popular support—73 to 76 percent (Tables 27 and 28). Analysis of public opinion along demographic and geographic lines reveals minor variation in the strength of support. More Easterners (81 percent) support the 55 mph speed limit than Westerners (72 percent), as do more females (84 percent) than males (66 percent). Although support declines as incomes increase, 67 percent of those in the highest income group support retention of the law. Support increases with the age of respondents, yet 72 percent of those under 30 years old support the law.

TABLE 27 Public Opinion of the 55 mph Speed Limit (*1*)

Question	Response (%)						
		1982	1981	1980	1979	1977	1974
Do you favor or oppose keeping the 55 mph speed limit on the nation's highways?	Favor	76	75	81	71	76	73
	Oppose	21	23	17	26	22	24
	No opinion	3	2	2	3	2	3

TABLE 28 Public Support of the 55 mph Speed Limit: Demographic Breakdowns, June 23–26, 1982 (*1*)

Category	Favor	Oppose	No Opinion
Sex			
Male	66	32	2
Female	84	15	1
Race			
White	74	24	2
Nonwhite	79	20	1
Education			
College	74	25	1
High school	74	24	2
Grade school	77	22	1
Region			
East	81	17	2
Midwest	75	24	1
South	70	28	2
West	72	25	3
Age			
Under 30	72	26	2
30 to 49	73	25	2
50 to 64	73	26	1
Over 65	88	10	2
Income ($)			
25,000 and above	67	31	2
20,000–24,999	72	28	–[a]
15,000–19,999	67	31	2
10,000–14,999	86	13	1
5,000–9,999	83	14	3
Less than 5,000	84	15	1

[a]Less than 1 percent.

PREFERRED SPEEDS

The question asked in the Gallup Poll does not address how the public would choose among alternatives; that is, what speed limits individuals would prefer, or whether maximum posted speeds should be set by the states rather than by the federal government. A few national polls have addressed these issues. In 1979 drivers were asked about their preference for a speed limit on major highways: 58 percent chose 55 mph, another 20 percent chose 60 mph (Table 29). When asked whether the maximum speed limit should be the prerogative of the federal government or the states, 68 percent preferred a national maximum (Table 30). Among those who preferred a state-established maximum, 54.4 percent preferred 55 mph or less. Responses to the preferred speed limit question varied

TABLE 29 Preferred Speed Limit, Drivers Only (2)

Question	Response (mph)	Drivers
Whether or not you yourself drive, in general, what do you think the speed limit for passenger cars should be on major highways?	Less than 55	4.6
	55	58.0
	56 to 60	20.2
	61 to 65	12.2
	66+	4.3
	Undecided	0.7
Total		100.0
Number	(1,323)	

along demographic lines; however, as with the Gallup Polls, the variation is only in the strength of support (Table 31).

OPPOSITION TO THE NATIONAL MAXIMUM SPEED LIMIT

The typical person opposed to the 55 mph speed limit is a male, aged 30 or less, living in the South or West, and with annual earnings above the national median income (Table 28). Another key indicator of opposition is the amount of driving done. Among those driving more than 30,000 miles each year, support for the 55 mph speed limit declines to only 53.7 percent (Table 32). Not surprisingly, these drivers report that their usual driving speed is much higher than those supporting the 55

TABLE 30 Preference for National or State Maximum Speed Limits (2)

			Preferred Speed Limit, by Choice, of Federal- or State-Set Maximum Limits		
Question	Response	1979	Preferred Speed	Federal Limit %	State Limit %
In your opinion, should the maximum speed limit be the same throughout the country, or should each state set its own speed limit?	Same	68.3	Less than 55	4.2	5.2
	State limits	30.2	55	62.0	49.2
	Don't know	1.5	56 to 60	18.8	23.2
			61 to 65	10.8	15.6
			66+	3.5	6.0
			Undecided	0.7	0.8
			TOTAL	100.0	100.0
			Number	(903)	(400)

TABLE 31 Preferred Speed Limit of Different Age Groups and Different Regions (2)

Preferred Speed Limit (mph)	Age: Less Than 30	Age: 30–44	Age: 45+	Region: NE	Region: S	Region: MW	Region: W
Less than 55	3.8	3.9	5.5	7.5	3.9	3.4	3.9
55	51.9	54.0	66.7	67.0	58.5	55.3	50.0
56 to 60	27.5	21.1	13.2	18.0	18.9	22.5	22.2
61 to 65	10.8	15.1	11.7	5.4	13.4	13.2	17.0
66+	5.3	5.0	2.7	1.4	4.8	5.6	4.8
Undecided	0.7	0.9	0.2	0.7	0.5	0.0	2.2
Total	100.0	100.0	100.0	100.0	100.0	100.0	100.0
Number	(397)	(430)	(454)	(294)	(434)	(356)	(230)

mph speed limit. This group, however, accounts for only 5 percent of those polled. Drivers whose usual driving speed is not more than 60 mph are solidly in favor of the speed limit. As usual driving speed increases, support for the speed limit declines (Table 33). Drivers opposed to the 55 mph speed limit are also more likely to be involved in accidents than supporters. The drivers who oppose the 55 mph speed limit have more accidents (Table 34). This increased accident frequency may be related to the preference for high speeds, which perhaps would indicate some causal relationship between speeding and accident frequency. Those opposed to the national maximum speed limit, who tend to be young males, are the most accident-prone group of all drivers. Drivers opposed to the 55 mph speed limit, however, also drive much more than the national average; hence their exposure to risk is greater and they are more likely to have been involved in an accident.

ATTITUDES TOWARD ENFORCEMENT

The surveys do not indicate increasing hostility toward the enforcement community or traffic laws. In 1980 48.3 percent of survey respondents indicated enforcement of the 55 mph speed limit was "about right," and 43.7 percent indicated it was not "strict enough." Only 4.3 percent indicated it was "too strict" (Tables 35 and 36). Interestingly enough, there is relatively little difference among drivers on this issue regardless of the amount of driving done each year.

Among the heaviest users of the highways, more drivers think that enforcement is "about right" and fewer think it should be stricter. Although more drivers do think that it is too strict, at 8 percent this is still a small percentage (Tables 35 and 36). In addition, a 1982 Gallup Poll reports

TABLE 32 Attitude Toward the 55 mph Speed Limit by Amount of Driving (*3*)

			Response							
			Strongly Favor		Somewhat Favor		Somewhat Opposed		Strongly Opposed	
Question	Miles Driven	(%)	N	(%)	N	(%)	N	(%)	N	(%)
How do you feel about maintaining the present 55 mph speed limit on the nation's highways?	0–2,000	18	160	74.4	31	14.4	12	5.6	9	4.2
	2,000–5,000	16	141	60.5	51	21.9	26	11.2	13	5.6
	5,000–10,000	23	196	60.5	64	19.8	34	10.5	27	8.3
	10,000–15,000	22	187	56.2	71	21.3	42	12.6	33	9.9
	15,000–20,000	9	78	52.0	40	26.7	15	10.0	17	11.3
	20,000–30,000	7	64	47.8	28	20.9	19	14.2	23	17.2
	More than 30,000	5	39	36.1	19	17.6	25	23.1	22	20.4

TABLE 33 Usual Driving Speed for Those Opposed to the 55 mph Speed Limit (*4*)

Usual Driving Speed	Drivers in Group		Percent Opposing 55 mph
	Number	Percentage	
55 or less	784	52.7	9.7
56–60	560	37.6	29.9
61–65	106	7.1	64.2
66–70	34	2.3	73.8
70 and above	5	0.3	80.0

TABLE 34 Reported Accidents Compared to Opposition to the 55 mph Speed Limit (*4*)

Number of Accidents Reportedly Involved in	Opposing the 55 mph	
	(%)	N
None	20.1	173
One	23.8	98
Two	29.7	41
Three or more	32.5	26

TABLE 35 Attitude Toward Enforcement (*3*)

Question	Response	1980	
		(%)	N
What is your opinion of current police enforcement of the 55 mph speed limit?	Too strict	4.3	64
	About right	48.4	723
	Not strict enough	43.6	654
	Don't know	2.7	41
	No opinion	1.0	15

TABLE 36 Attitude Toward Enforcement by Amount of Driving (*3*)

Miles Driven Last Year	Too Strict		About Right		Not Strict Enough		Don't Know		No	
	N	(%)	N	(%)	N	(%)	N	(%)	N	(%)
0–2,000	7	3.3	93	43.3	99	46.0	11	5.1	5	2.3
2,000–5,000	10	4.3	106	45.5	109	46.8	6	2.6	2	0.9
5,000–10,000	8	2.5	161	49.7	143	44.1	10	3.1	2	0.6
10,001–15,000	11	3.3	163	48.9	146	43.8	9	2.7	4	1.2
15,001–20,000	9	6.0	74	49.3	65	43.3	1	0.7	1	0.7
20,001–30,000	10	7.5	69	51.5	51	38.1	3	2.2	1	0.7
More than 30,000	9	8.3	57	52.8	41	38.0	1	0.9	0	0.0

TABLE 37 Support for the 55 mph Speed Limit and Frequency of Being Stopped by Police (*1*)

Question	Response	1982 (%)	1981 (%)
Have you been stopped by the police for exceeding the 55 mph speed limit?	Yes	29	19
	No	71	81
Do you favor or oppose keeping the 55 mph speed limit on the highways of the nation?	Favor	76	75
	Oppose	21	23
	No opinion	3	2

a consistent level of support for the 55 mph speed limit despite the increase in the percentage of those polled who have been stopped for speeding. Although only 19 percent of those polled had been stopped for speeding in 1981, 29 percent had been stopped by 1982. This increase, however, has not affected overall support for the law (Table 37).

ACTUAL DRIVING BEHAVIOR

Although support for the law should make enforcement easy, enforcement of the compliance standard (50 percent of state traffic in compliance with the law) is difficult because most drivers exceed the speed limit, especially on the Interstates. Even most of the supporters interviewed in national polls from 1978 to 1980 admitted that they did not strictly comply with the posted speed (Table 38). In a 1979 poll the relationship between actual driving speeds and preferred speed limits was examined, and the results revealed that although 58 percent actually

TABLE 38 Preferred Speed Limit and Normal Driving Speed (*3*)

Question	Response	1978	1979	1980
On roads where the speed limit is 55 miles per hour, how fast do you usually drive?	55 or less	50.0	52.4	54.8
	56 to 60	40.0	37.4	36.6
	61 to 65	7.7	7.1	6.8
	66 to 70	1.7	2.3	1.4
	Over 70	0.5	0.3	0.4
How do you feel about keeping the 55 mph speed limit?	Strongly favor	55.5	57.5	57.7
	Somewhat favor	20.8	19.4	20.3
	Somewhat opposed	13.4	12.5	11.6
	Strongly opposed	9.7	10.1	9.6
	No opinion	—	—	0.7

TABLE 39 Preferred Speed Limit and Typical Driving Speed (*2*)

Question	Responses (mph)	Drivers (a) Preferred Speed	(b) Typical Driving Speed
(a) What do you think the speed limit for passenger cars should be on major highways? (1979 survey)	Less than 55	4.6	11.0
	55	58.0	37.4
	56 to 60	20.2	37.2
	61 to 65	12.2	11.6
(b) Suppose you're on a highway with a 55 mph speed limit, that it is daytime, the weather is good and traffic is moderate. In that case, what is the actual speed you normally find yourself driving at?	66+	4.3	2.2
	Undecided	0.7	0.4
	Number	(1,323)	(1,323)

preferred a 55 mph speed limit, only 37 percent said they normally drove 55 mph (Table 39).

The apparent incongruity between support for a 55 mph speed limit and actual driving results from the widespread public perception that driving a few miles per hour faster than the posted speed is in compliance with the spirit of the law. Most drivers appear to believe that speeds up to 5 mph in excess of the posted speed limit should be tolerated. Although more than 70 percent support the 55 mph speed limit, only 18 percent believe that the police should issue citations at the point when speeds exceed 55 mph (Table 40). The majority—75 percent—did not believe that officers should begin issuing citations until speeds exceed 60 mph. Support for the 55 mph speed limit is clearly conditioned by the public's expectation that they can drive 60 mph and not be cited for speeding.

Because most drivers believe that they comply with the law if they drive no more than 60 mph, the data that indicate the majority of traffic exceeding 55 mph on major highways appear in a new light. In 1979

TABLE 40 Expected Tolerance in Enforcement (*4*)

Question	Response (mph)	1978	1977
At what speed do you think patrolmen should begin to give speeding tickets where 55 is the speed limit?	Over 55	18.5	15.3
	Over 58	9.4	9.4
	Over 60	47.5	51.6
	Over 65	19.0	17.0
	Over 70	2.6	4.0
	Don't know	2.9	2.7

when 75 percent of the public replied to the pollsters that they expected a 5 mph tolerance, about this same percentage of drivers on the rural Interstates (68 percent) drove within the tolerance range, although 72 percent were violating the letter of the law.

Another explanation for the discrepancy between actual driving speeds and the national polls of driving preferences results from the difference in the groups being compared. The national public opinion polls sample all drivers, whereas the speed data measures drivers on the road. Although these two groups overlap, the strongest support for the law comes from those who tend to drive the least. Among the heaviest users of the highways—persons driving more than 30,000 miles each year—support for the speed limit drops to 53.7 percent (Table 32). In contrast, public opinion polls, which record each individual's opinion regardless of the amount of driving, do not reflect the actual use.

SUMMARY

During the last decade the majority of Americans has consistently supported the 55 mph speed limit. Why then does this support apparently disappear when these same Americans take to the highways? The answer to this question involves at least two subtleties. First, even most supporters of the 55 mph speed limit believe that driving at 60 mph is, in effect, in compliance with the spirit of the law.

The second subtlety emanates from the difference between polls of adults and observations of vehicle speeds. The strongest supporters of the law drive the least; 89 percent of those driving less than 2,000 miles each year support the law. But these same supporters—because they drive so little—are the least likely to be out on the road. The strongest opponents drive more than 30,000 miles each year. Although they make up a relatively small percentage of those polled, by driving more they make up a much larger percentage of total traffic, thereby making their opposition more evident.

Public opinion polls reveal that regional distinctions exist, but also indicate overwhelming support for the law in all areas of the country. Given that traffic enforcement agencies provide more leeway before issuing citations in some areas of the country than in others (as discussed in Chapter 9 and Appendix F), support for the law across regions probably reflects these varied levels of enforcement stringency.

Although the 55 mph speed limit places more demands on enforcement agencies—more motorists exceed the current speed limit than exceeded state-set maximums—the polls indicate no evidence of increased hostility toward the police. Even though more motorists indicated they had been stopped for speeding in 1982, results of the Gallup Poll continue to indicate strong support for the law. The polls also indicate that most people do not object to having the federal government, rather than the states, set the maximum speed limit.

REFERENCES

1. Gallup Poll 1982. *The Gallup Report: Political, Social, and Economic Trends*. Report 205. Princeton, N.J.
2. A. Vayda and I. Crespi. *Public Acceptability of Highway Safety Countermeasures*. DOT HS-805-971. Safe Driving Conformance Research, Vol. II, Mathmatica Policy Research, Inc., Princeton, N.J., June 1981.
3. R.L. Berger and G.S. Persinger. *Survey of Public Perceptions on Highway Safety, 1980*. HS-805-702. NHTSA, U.S. Department of Transportation, 1980.
4. Teknekron Research, Inc. *1979 Survey of Public Perceptions on Highway Safety*, NHTSA, U.S. Department of Transportation, July 1979.

9
Enforcement and Compliance

Before 1974 state and local governments shared responsibility for setting and enforcing speed limits. With the establishment of the 55 mph maximum speed limit the federal government became directly involved in regulating traffic speeds by setting the maximum permissible speed. Enforcement of speed limits continues to be a state and local responsibility. Local governments continue to have primary responsibility for enforcing the speed limit within their own jurisdictions. State agencies share responsibilities on state highways that traverse local government lines and have primary responsibility for state highways in rural areas.

Before the establishment of a maximum speed limit, state and local traffic engineers relied on the 85th percentile rule for posting speed limits. This ensured that the majority of motorists were automatically in compliance. By contrast, the 55 mph maximum speed limit attempts to regulate the behavior of the average motorist.

Upon making the 55 mph speed limit a permanent safety policy in 1975, Congress required the governors of each state to certify that their states were making a sufficient effort to enforce the law and required state reports on speed trends on highways posted at 55 mph. Under this legislation, however, the federal government had little power to influence the enforcement efforts of individual states. With the passage of the Highway Safety Act of 1978, the states were threatened with loss of highway funds if they did not maintain a set percentage of traffic (currently 50 percent) in compliance with the maximum speed limit.

TABLE 41 Percent of Traffic Exceeding 55 mph (average of states, unadjusted data) (*1*)

Highway System	1980	1981	1982	1983	Percent Change 1982–1983[a]
Urban Interstate	51.2	51.0	58.4	60.5	3.6
Rural Interstate	65.9	67.6	73.1	73.6	0.7
Other freeways	–[b]	47.1	53.8	56.6	5.2
Urban arterials	–[b]	30.2	32.4	34.3	5.8
Rural arterials	–[b]	43.2	46.2	47.5	2.8
Rural collectors	–[b]	32.7	34.8	36.6	5.2
All statewide highways polled at 55 mph	49.0	48.5	52.6	54.1	2.8

[a]Percentage change calculated for 1982–1983 because these are the only 2 years with consistent speed measurement in all states.
[b]Data not available.

Despite these regulations and the increased enforcement efforts of many states, fewer drivers are complying with the 55 mph speed limit than at the time these amendments were enacted into law in 1978. More motorists are speeding on almost all highway systems, although the largest increases in noncompliance are occurring on urban highways and rural two-lane collectors (Table 41). The figures given in Table 41 are based on the raw speed data submitted by the states to the federal government. The states are allowed to adjust these data for potential errors caused by inaccurate speedometers and for sampling error. As demonstrated in the second half of this chapter, these adjustments overstate the error in the data, and for statistical analyses the raw compliance data are a better representation of motorist behavior. The actual percentage of drivers speeding, if not for the adjustments allowed to the data, would place most states in jeopardy of being sanctioned. On the basis of 1983 compliance reports, 37 states exceeded the standard before their compliance data were adjusted, an increase from just 30 states the previous year (Figure 25). The number of states with more than 50 percent of traffic exceeding 55 mph (unadjusted data) is given below (*1*).

Year	*No. of States*
1980	23
1981	21
1982	30
1983	37

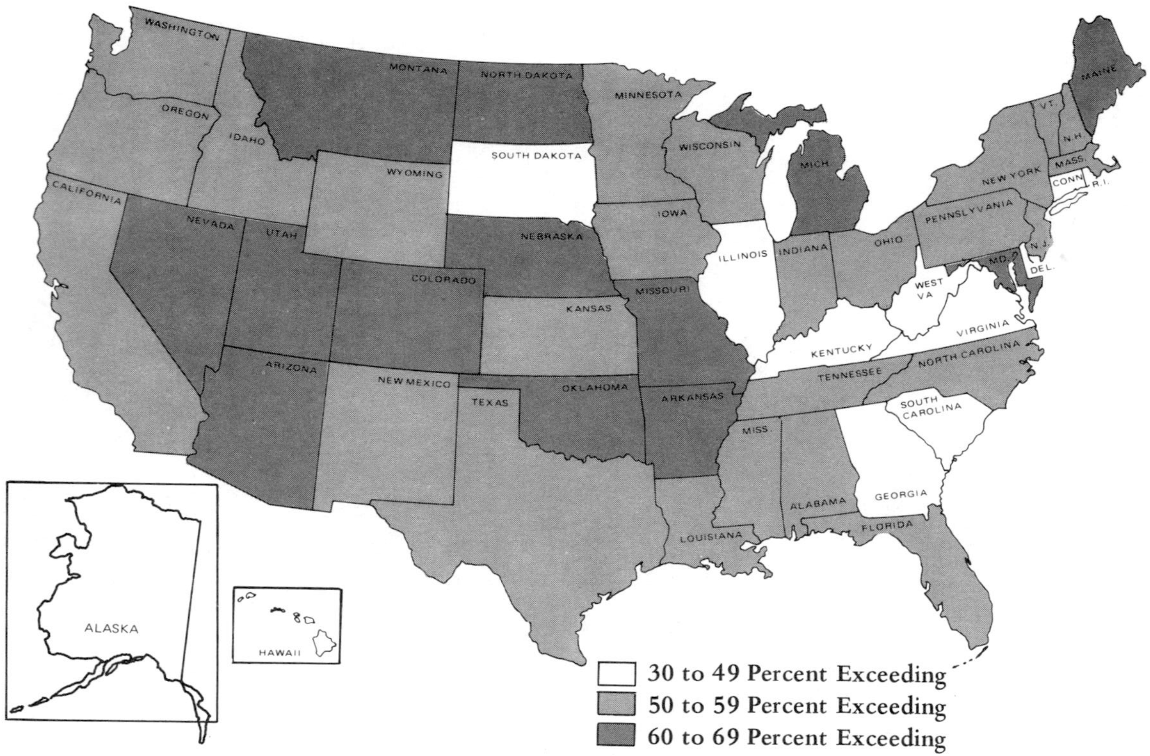

FIGURE 25 Percentage of motor vehicles exceeding 55 mph on public highways, 1983 (unadjusted data) (*1*).

During this same period of eroding compliance, a number of states weakened their penalties for violations of the 55 mph speed limit. The Congress, concerned about the enforcement efforts of these states, requested that an examination of "whether the laws of each state constitute a substantial deterrent to violations of the maximum speed limit" be undertaken by this study.

Motorist compliance with the 55 mph speed limit is somewhat worse than average in most of the states with weakened penalties, but there is little basis for inferring a cause-and-effect relationship between statutory penalty levels and the amount of speeding. The mechanism for deterrence of speeding involves far more than simply the penalties established in state law. Deterrence also depends on the actions of state and local enforcement agencies and traffic courts and is influenced by public acceptance of the law. In the following section these aspects are examined to address the question of whether the laws of the states constitute a substantial deterrent. Then the federal government's interaction with this involved compliance mechanism is reviewed. The chapter concludes with suggestions for improving federal provisions for speed limit enforcement.

THE DETERRENCE MECHANISM

The mechanism by which motorists are deterred from violating the 55 mph speed limit involves the institutions that make, enforce, and adjudicate traffic laws, as well as the perceptions of individuals of the appropriateness of the law and the risks involved in its violation. Those institutions responsible for establishing and enforcing the law both influence, and are influenced by, public acceptance of the law. Although the general public usually complies with laws that it accepts, unpopular laws are sometimes quickly overturned. The U.S. Congress's reversal on the interlock safety-belt requirements in 1975 is one example.

The deterrence mechanism also depends on the perceptions of the motorist on the road. The decision of a driver to abide by a traffic law involves a judgment about the appropriateness of the intended regulation. Most drivers recognize that traffic laws are designed to regulate the safe flow of traffic rather than to prohibit behavior that is morally wrong (*2*). Drivers regularly exceed the posted speed limits without thinking of themselves as criminals. The extent to which they do so apparently reflects their expectations about the stringency with which the law will be enforced, the risk of being stopped and fined, and the risk to their own safety.

All of these factors—institutional interactions and public perceptions—influence motorist compliance. Before addressing the question of how federal laws can most productively influence compliance, it is useful to examine four elements that influence compliance and review experience with each: (a) state laws, (b) traffic courts, (c) enforcement, and (d) public acceptance.

State Laws

The penalty imposed in state laws is popularly regarded as the key deterrent to violations of the law. The general public appears to believe that increasing legal sanctions will reduce crime rates despite the lack of evidence that such policies work (*3*). Experience with past crackdowns on speeding has shown that an overreliance on the severity of the penalty has the adverse result of decreasing the certainty that it will be applied.

In 1955 the state of Connecticut dramatically increased the penalty for speeding because of a rapid increase in highway fatalities in the preceding year. Drivers convicted of speeding were subject to loss of their license for 30 days on the first conviction. In addition, the governor required vigorous enforcement of the law. These changes received wide notice in the press. In the months immediately following announcement of the new policy, the state police reported that the number of speeders had declined to only 2 percent of all drivers. These benefits were short lived, however; in subsequent months the judicial system, and perhaps also the police, proved "unwilling to invoke penalties that might seem severe and unfamiliar in context" (*4*).

Research into the effectiveness of increased severity on crime rates in general suggests that increasing the severity of penalties can undermine the certainty that they will be implemented. As penalties become more severe, those accused have more incentive to fight their cases, which, in turn, slows down the judicial process (*3*). The lack of certainty of the penalty may diminish the perception of risk in the minds of potential lawbreakers.

In the last few years the trend toward reduced penalties for speed-limit violations has led Congress to question whether the penalties in these laws are sufficiently steep to deter speeding. These concerns are reflected in the request for this study. A few states have passed laws that appear to be specifically aimed at vitiating the 55 mph speed limit (Table 42). Nevada, the best known example, passed a $5 energy-wasting fee for violations of the speed limit up to 70 mph (the total penalty is $15 with court costs). Oklahoma and South Dakota have passed laws that shield a driver's record from public scrutiny for speeding violations below the pre-1974 maximum speed limit. This prevents drivers from being penalized by their insurance companies for speeding violations below the previous maximum limit (state laws are summarized in Appendix D).

In addition to the penalties established in state laws, a convicted speeder must also pay court costs. Although data on court costs imposed by traffic judges are not collected by the states, in some jurisdictions the court costs equal or exceed the fine.

Probably the highest financial penalty associated with a speeding conviction is an increase in insurance rates. These increases are triggered by the number of points assessed against the driver's record when convicted. Slightly more than one-half of the states have some type of point system. Typically, the number of points assessed increases with the speed above the posted limit.

TABLE 42 States With Weakened Penalties for Violations of the 55 mph Speed Limit (5)

State	Previous Penalty	Current Penalty
Arizona	$100 maximum fine	$15 for violating 55 mph national maximum speed limit between 56 and 65 mph
Iowa	$20 for speeding at 65 mph	$10 fine for speeding between 56 and 65
Kentucky	$10 to $100 fine, points assessed	No points assessed for speeds less than 70 mph on Interstate
Nebraska	$10 to $100 fine, points assessed	$10 fine for speeding between 56 and 65; no points assessed or points assessed on Interstates
New Mexico	$200 maximum plus points	$1 per mph between 56 and 70; no points assessed
North Dakota	65 mph fine equals $20 and 2 points	65 mph fine $10 and one point
Nevada	$500 maximum fine plus points	$15 energy-wasting fee up to previous speed limit
Oklahoma	$10 to $200 fine	Same fine; violation for 55 mph national maximum speed limit not placed on drivers record
Oregon	$100 maximum fine	Fine the same but cannot lose license for accumulation of 55 mph violations only
South Dakota	$100 maximum fine, no points assessed	Same maximum fine; 65 mph violations as low as $20 with 2 points; No public access to records of violations of 55 mph national maximum speed limit

NOTE: See Appendix D for details on other state penalties.

Thus, the total penalty for speeding varies widely across the states depending on the level of the fine, the court costs imposed, and the impact of a state's point system on a driver's insurance rates. Comparable data on these costs are not available. Despite this inability to distinguish precisely between the actual penalties imposed by state laws, the laws themselves reveal that the maximum penalties in some states are well below those of other states.

A number of states impose modest fines for minor violations of the 55 mph speed limit. Fines in Arizona are about $15 and no points are assessed against a driver's record for speeding violations of less than 65 mph. In Iowa speeding violations of less than 61 mph are not entered on the driver's record. In Nebraska and Kentucky, fines for violations of the speed limit on the Interstate system are lower than fines for the same violations on other highways, and points are not assessed for speeds lower than the state limit that prevailed before the national maximum speed limit. As noted previously, some states do not provide insurance companies access to drivers' records for violations below the pre-55 mph state speed limit.

To compare the effects of state laws, three groups of states may be identified based on the relative level of penalty imposed for speeding (Table 43):

1. *Higher-penalty states* impose points for speeding violations between 55 mph and 70 mph (or the previous state maximum), and impose maximum fines greater than $25 (there are 26 such states);

TABLE 43 States With Higher and Lower Penalties for Violations of the 55 mph Speed Limit (*1*)

Higher-Penalty States (N = 26)		Medium-Penalty States (N = 17)		Lower-Penalty States (N = 7)
Alabama	Maine	Connecticut	Ohio	Arizona
Alaska	Maryland	Iowa	Oklahoma	Idaho
Arkansas	Nebraska	Kansas	Oregon	Massachusetts
California	New Jersey	Michigan	South Dakota	Montana
Colorado	New York	Minnesota	Tennessee	Nevada
Delaware	North Carolina	Mississippi	Texas	New Mexico
Florida	Pennsylvania	Missouri	Washington	Rhode Island
Georgia	South Carolina	New Hampshire	Wyoming	
Hawaii	Utah	North Dakota		
Illinois	Vermont			
Indiana	Virginia			
Kentucky	West Virginia			
Louisiana	Wisconsin			

2. *Medium-penalty states* impose one of two measures: points for speeding between 55 and 70 mph or maximum fines greater than $25 (1 state meets the first criterion and 16 meet the second);

3. *Lower-penalty states* do not impose points for speeding violations between 55 and 70 mph and have maximum fines of $25 or less (7 states are in this category).

The low-penalty states report substantially higher speeds (Table 44). In 1983 these states averaged 58.5 percent of motorists exceeding the posted speed limit, as compared to an average of 51.9 percent of motorists in the states with stronger penalties. The states with low penalties also have a higher fatality rate on highways posted at 55 mph [3.14 fatalities per 100 million vehicle miles as compared to 2.76 in states with higher penalties (Table 44)].

These averages, however, conceal important exceptions. For example, six high-penalty states have a noncompliance level above the average of the low-penalty states (Table 45). In addition, 1 of the weak-penalty states, Idaho, has a compliance level better than 14 of the states in the high-penalty group. Between 1980 and the present the trends in increased speeding are the same in all three groups. More motorists are driving faster regardless of the level of the penalty imposed.

At first glance the comparison between compliance and state penalties suggests a relationship between the amount of speeding and the penalty imposed. However, any interpretation of this relationship is complicated by differences in the roads posted at 55 mph, by variations in the frequency

TABLE 44 Trends in Compliance, Frequency of Citations, and Fatality Rates for States With Stronger and Weaker Penalties (*1*)

Category	Percentage of State Traffic Exceeding 55 mph				Percentage Change	Citation Rate (per million vehicle miles)	Fatality Rate (per million vehicle miles)
	1980	1981	1982	1983	1980–1983[a]	1982	1982
Higher-penalty states (N = 26)	46.7	47.7	49.9	51.9	+11.1	9.78	2.76
Medium-penalty states (N = 17)	51.0	48.1	54.1	56.3	+10.4	11.10	3.1
Lower-penalty states (N = 7)	54.9	52.4	61.4	58.5	+ 6.6	20.92	3.14

[a] Because of changes in speed monitoring techniques, some of the increases between 1981 and 1982 are attributable to changes in measurement.

TABLE 45 States With High and Low Penalties Rank Ordered by Level of Compliance (unadjusted data, 1983) (*1*)

Percent Exceeding 55 mph 1983	High-Penalty States	Medium-Penalty States	Low-Penalty States
69.0	—	North Dakota	—
67.4	Nebraska	—	—
66.3	—	Michigan	—
63.7	Maryland	—	—
63.6	—	Wyoming	—
63.5	—	—	Nevada
62.6	—	Missouri	—
62.3	—	—	Arizona
61.6	Colorado	—	Montana
60.8	Maine	—	—
60.7	Utah	—	—
60.4	—	Oklahoma	—
59.6	Vermont	—	—
57.9	Alabama	—	—
57.7	—	Kansas	—
57.6	Florida	—	—
57.3	—	New Hampshire	—
56.9	Delaware	—	—
56.8	California	—	—
56.6	—	—	Massachusetts
56.6	—	—	Rhode Island
56.3	New York	—	New Mexico
56.0	—	Minnesota	—
55.8	—	Texas	—
55.7	—	Iowa	—
55.5	—	Ohio	—
53.9	Indiana	—	—
53.6	Wisconsin	Washington	—
53.0	—	Tennessee	—
52.5	—	Oregon	—
52.4	Pennsylvania	—	—
52.3	North Carolina	—	Idaho
51.8	—	Mississippi	—
51.0	New Jersey	—	—
50.7	Louisiana	—	—
49.4	—	Connecticut	—
49.2	Georgia	—	—
49.2	Illinois	—	—
45.4	—	South Dakota	—
44.9	South Carolina	—	—
43.4	Arkansas	—	—
42.6	Virginia	—	—
39.9	Kentucky	—	—
36.7	Hawaii	—	—
36.6	West Virginia	—	—
33.9	Alaska	—	—

of citations, and by the possibility that some third factor, for example, public opinion, is affecting both the laws and the degree of compliance.

Different Roads

Some states post only high-quality Interstate highways at 55 mph, whereas other states have hundreds of miles of two-lane rural roadways posted at 55. When the states are compared on the basis of compliance on their Interstates alone, the apparent relationship between state laws and compliance rates breaks down. Both the strong-penalty states and the weak-penalty states report that 73 percent of motorists are traveling faster than 55 mph on rural Interstates.

Frequency of Citations

Surprisingly, the low-penalty states demonstrate a higher level of enforcement activity per vehicle mile traveled on highways posted at 55 mph. The low-penalty states cite speeders twice as frequently as the high-penalty states. Perhaps more citations are issued by low-penalty states because they have more difficulty meeting the federal compliance standard; but in any case it appears that they offset their weaker penalties, at least to some extent, by more vigorous enforcement.

Other Causes

Athough low-penalty states report, on average, higher speeds and higher fatality rates than other states, this may not reflect a link between the laws and motorist behavior. Instead, both may be a reflection of some other cause; for example, public attitudes. Lower penalties and faster driving speeds may both be in consonance with dominant state cultural norms. Additionally, these categorizations based on the laws on the books may bear little relationship to how these laws are actually interpreted by individual traffic judges. Traffic courts are given wide discretion in most states in applying the penalties, and states with high maximum fines may well have average fines no higher than those states with modest penalties.

Traffic Courts

Untangling the relationship between the laws of each state and their effect on compliance levels is stymied by a number of problems, not the least of which is a lack of information about the penalties that convicted speeders actually receive. The majority of state laws prohibiting speeding, like other traffic violations, establish a range of penalties that traffic judges apply at their discretion. Some states allow judges to postpone or set aside judgments. For these reasons, the penalty for speeding tends to vary with individual traffic judges regardless of the state law, and the penalty may vary as much within individual states as it does among them collectively. Generally, however, the actual fines imposed for exceeding 55 mph are unknown, because states do not compile information about the fines handed out in individual courts.

Some evidence suggests that the average penalty imposed in traffic courts is well below the maximum. For example, Georgia has one of the highest maximum fines on the books ($1,000), but the average fine imposed is only $40 (*5*). Indiana and Florida have a maximum fine of $500, but again the average fine is much lower, about $25. Although Maine has a lower maximum fine than these states, traffic judges in the state fine speeders an average of $40.

Except in these four states where data on average fines are routinely collected, data on average fines do not exist. Although a survey of fines actually imposed could be conducted, the sample would have be to be extremely large to account for the expected variation within the states as well as the differences among them. This would be a sizable research task, well beyond the scope of this study (Appendix H).

Not surprisingly, drivers themselves are confused about actual penalties for speeding. A 1977 poll of California motorists revealed that 39 percent had no opinion about whether the existing penalty for speeding was too strict, probably because 23 percent admitted to having no knowledge of the penalty (*6*). Another 10 percent of those polled guessed that the fine was less than $20 and an equal percent guessed that it was more than $40. Some 28 percent guessed that it depended on the speed in excess of the limit. At that time the maximum fine for speeding in California was $50, and the actual fine depended on the traffic judge. Unless most drivers are aware of the penalty involved, increasing or decreasing the penalty is unlikely to have much effect on overall speeding.

Enforcement

The perceived risk of being apprehended for a speeding violation, whether the risk is real or imagined, may be the key factor in deterrence. Motorists probably respond more from the simple fear of being stopped for speeding than from the fear of the actual penalty involved (*7*). Most drivers intuitively realize that the risk of being ticketed for speeding is extremely small, because many regularly speed without being stopped. A change in their behavior requires that they believe they run an increased risk of being stopped.

A few states have experimented with intensive enforcement of 55 mph speed-limit violations, and these efforts have had a measurable impact on motorist behavior. For example, California increased enforcement on five heavily traveled Interstates in 1982 by allocating an additional $1.3 million for overtime enforcement. One route was later dropped from the program, and on the remainder some 40,000 patrols were deployed. During weekdays—when the effort was underway—overall compliance improved 5 percent (from 78.2 percent exceeding 55 mph to 74.7 percent). On weekends, when the patrols were not deployed, speeding actually increased by 10 percent (*8*).

Similarly, the New York State police increased their enforcement efforts on 600 miles of Interstate to test the efficacy of different patrol configurations. Speeds were reduced during the program and remained low regardless of enforcement strategy (*9*). During the effort average speeds were reduced to about 6 percent below the average rural Interstate speeds for the same period. Similar enforcement crackdowns in Utah and Connecticut revealed that as the number of patrol officers increased per unit of highway, the percent of traffic exceeding the speed limit declined. However, when the patrolmen were removed, the reductions in speed disappeared (*10*).

Similar results have been demonstrated in many other evaluations of speed-limit enforcement (*7,11–17*). These same studies also demonstrate that the effects of enforcement diminish rapidly when the patrols are removed, usually completely within 1 or 2 days. Although these tests of various forms of stepped-up enforcement demonstrate that traffic can be slowed down when patrolmen are visible and active, little is revealed about the practical extent to which statewide enforcement efforts shape the speeds reported on all state highways. Because there are 190 miles of 55-mph highway for every on-duty patrolman, any realistic enforcement effort will leave large sections of road unmonitored (Appendix F).

Nationwide, enforcement of the 55 mph speed limit increased immediately after passage of the law in 1974 as evidenced by the increased number of citations in 1975. Although the number of citations issued has increased above the level issued in 1975, this number has not quite kept up with the increased amount of traffic on the road, and the actual number of citations issued declined in 1982 and 1983. The number of citations issued for speeding violations is probably a poor measure of enforcement, however. A comparison of speed control strategies in

Texas has shown that the visibility of patrolmen has a direct effect on vehicle speeds, whereas the frequency of citations shows no relationship to speeds (7). Nevertheless, the number of citations issued each year is the only measure of enforcement that the federal government requests of the states as part of its annual compliance certification. Accordingly, these data provide about the only systematic basis for comparison, as examined next.

A comparison of 1982 data on state speeding and the frequency with which citations are issued reveals a statistically significant correlation, but in a surprising direction. Instead of demonstrating that the states that issue fewer citations also have more speeding, the correlation indicates that the states that issue citations most frequently also have the most speeding (rho = 0.47, significant at the 0.01 level).

However, when the change in compliance from one year to the next on all state highways posted at 55 mph is compared with the changes in citation frequency, no relationship can be determined. The changes in citations and compliance between 1980 and 1981 and between 1981 and 1982 indicate no correlation at the state level. These separate approaches to analyzing the relationship between frequency of citations and motorist compliance produce confusing and conflicting results. Comparison of state compliance for a single year reveals that in the states with the greatest compliance problem more citations are issued, which appears to indicate that issuing citations does not deter speeding.

In contrast, changes in compliance and citation frequency over time suggest no relationship exists at all. These contradictory results probably attest more to the limitation of citation frequency as a measure of enforcement than anything else. Careful studies of enforcement efforts (cited previously) convincingly demonstrate that intensive patrol efforts can slow traffic. Apparently the measure of frequency of citations is a poor substitute for the real effect of patrols on motorist behavior, and for this reason it is an inadequate basis for comparing state enforcement efforts.

State and Local Responsibilities

Any attempt to relate statewide enforcement to statewide compliance is also complicated by the proportion of speed-limit enforcement by county police, sheriff departments, or municipal police. Roughly one-fourth of all travel on highways posted at 55 mph occurs in urban areas in which local agencies have primary or concurrent responsibility for speed-limit enforcement. Jurisdictional responsibility varies widely by state.

Although in most states the state patrol or the state police are responsible for at least 75 percent of the enforcement of the 55 mph speed limit, in a few states enforcement is shared equally by state and local police; for example, New York and Arizona (*18*). In most states the state patrol has primary responsibility on main highways, such as the Interstates. Even on the Interstates, however, local police often share responsibility (and in some cases have primary responsibility) on the urban segments. Local agencies, however, do not always have the same priorities as the state. For example, in Michigan the sheriff's departments share responsibility for enforcement of the 55 mph speed limit on urban Interstates; however, the sheriff's department in the state's most populous urban area has no active patrols for enforcement of the 55 mph speed limit (Appendix F).

Public Acceptance

As noted in Chapter 8, public support for the 55 mph speed limit has remained strong since it became law in 1974. Polls reflecting public support, however, also reveal that acceptance of the law is conditioned by the expectation that some tolerance will be granted in its enforcement. Most drivers appear to believe that speeds up to 5 mph in excess of the posted limit should be tolerated by the police. Although more than 70 percent support the current maximum speed limit, only 25 percent believe that the police should issue tickets at the point when speeds exceed 55 mph (*19*).

Although tolerance levels vary from place to place, they reflect a balance between the courts, enforcement agencies, and public opinion. Small violations of the speed limit are difficult to measure and defend in court. Many traffic judges will not convict unless the violation is 5, and sometimes 10, mph above the limit. Judges contend that small violations clog already overburdened courts and divert attention from more serious offenses. One state patrol reports that prosecutors will dismiss cases unless the offender has been clocked at 10 mph above the speed limit (*18*). Presumably, this judicial behavior reflects prevailing notions about public acceptance and about the importance of different types of illegal behavior.

Patrolmen have come to know the type of violation that will result in conviction. Many enforcement agencies have written policies, usually allowing 5, and sometimes 10, mph above the posted speed before issuing a citation (Appendix F). Enforcement tolerances may have been modified somewhat by the 55 mph speed limit, but enforcement agencies have granted tolerances for the enforcement of all speed limits.

Differences in regional public opinion may partly contribute to the variations in enforcement of the 55 mph speed limit in various regions of the country. The price of public approval in different states or regions may require different strictness of interpretation. Thus the greater tolerances granted in western and plains states may reflect a workable balance between public opinon, law enforcement, and court policy that differs from the balance struck in eastern states.

In addition, public acceptance is undoubtedly linked to public awareness of the rationale for the law. For example, during the fuel shortages that followed the Arab oil embargo, motorists were keenly aware of the difficulty of getting fuel and the need to conserve it. Exceeding the 55 mph speed limit was not only perceived as wasteful but also as unpatriotic. Hence, some drivers honked at cars that sped past at speeds above 55 mph, and some obstinately drove 55 in the passing lane of multilane highways to discourage faster driving. Public support of this type altered the normal balance between enforcement and judicial and public behavior.

Public awareness of the safety benefits undoubtedly influences public acceptance of the law, although not with the same crisis mentality as the oil embargo. Public acceptance is also partly conditioned by the media attention that is given to the law and its supporting rationale.

Many of the elements that shape public acceptance of speed limits cannot be easily regulated. Public awareness of the benefits, mobilization to meet a crisis, increased awareness as a result of media campaigns, and many other uncontrollable forces shape public acceptance, and this eventually influences the speeds at which individuals will drive.

Before exploring the question of whether there are additional steps that the federal government should take to increase compliance with the 55 mph speed limit, it is instructive to review in some detail how the federal sanctions process now operates. In the next section, this process is described and its rationale and performance are critiqued. In the concluding section, several ways are suggested for making the process more realistic and effective, taking into account the complicated mechanics of deterrence that have been reviewed.

FEDERAL COMPLIANCE PROCESS

When the 55 mph speed limit was made permanent in 1975, the only federal provision for its enforcement was the requirement of an annual certification by the governors that their states were adequately enforcing the law. It soon became apparent that this was insufficient to ensure compliance in all states. The Congress perceived that some states were insufficiently enforcing the law, and, in response, enacted legislation in 1978 that required a specific measure of compliance based on the percentage of traffic exceeding the speed limit on all highways posted at 55 mph (*20*). For those states that failed to comply, some federal-aid highway funds would be withheld until they complied. The initial standard for 1979 was quite liberal, allowing as much as 70 percent of traffic to exceed the speed limit.

For subsequent years, Congress set a schedule that imposed a more stringent test each year. By 1983 no more than 30 percent of traffic could exceed the legal speed limit (Table 46). In later legislation, Congress eased the scheduled standards set out in 1978 (*21*). The law now requires that for 1981 and thereafter, 50 percent of traffic must be in compliance with the speed limit, and stipulates that a state not meeting this standard may forfeit up to 10 percent of its primary, secondary, and urban highway funds.

The states must also report the percentage of traffic exceeding 55 mph based on a formula established by the Secretary of the U.S. Department of Transportation. The law instructed that the Secretary, in developing this formula, should include "criteria which takes into account the variability of speedometer readings and criteria based upon speeds of all vehicles or a representative sample of vehicles" (*22*). This language provides the basis for the current adjustments outlined in the following sections.

TABLE 46 Standards and Sanctions in 1978 Legislation and Amended Standards and Sanctions in 1981 Legislation (*20,21*)

Year	If Statewide Percent Exceeding 55 mph Exceeds (−%)	State Can Lose Up To (−%) of Highway Aid
1979	70	5
1980	60	5
1981	50	5
1982	40	10
1983	30	10
Amended Standards and Sanctions in 1981 Legislation		
1981 and following	50	10

NOTE: The funds subject to penalty include primary, secondary, and urban highway aid. Funding for other highways, such as Interstates, are not affected.

Adjustments Allowed to State Data

As directed by law in 1980 the Federal Highway Administration (FHWA) established adjustments that the states could use to lower their estimates of the percentage of statewide traffic exceeding the 55 mph speed limit. The FHWA allowed adjustments to the speed data to account for errors caused by inaccurate speedometers, sampling, and inaccurate measures of vehicle speeds by monitoring devices. Other adjustments might also be accepted if sufficiently well documented.

In response to the adjustments suggested by the FHWA, California offered an alternate method that consolidated these adjustments into a single formula, which, when applied to its own 1980 compliance data, lowered the estimate of traffic exceeding 55 mph from 61.7 to 44.5 percent. This formula, known as the simplified method, was allowed by FHWA and has been adopted by most of the states.

The formula is as follows:

$$D = 0.5\,(A - B) + B$$

where A is percent exceeding 55 mph, B is percent exceeding 60 mph, and D is adjusted percent exceeding.

Example:

$$A = 61.7 \text{ percent}$$

$$B = 27.4 \text{ percent}$$

$$D = 0.5\,(61.7 - 27.4) + 27.4$$

$$= 44.5$$

The California adjustment, which lowered the apparent noncompliance by 17 percentage points, is fairly representative of the size of the adjustment most states receive from the simplified method. The average adjustment, if applied to the unadjusted 1982 data, is about 14 percentage points.

Speedometer Error

The adjustments to state compliance data to account for speedometer error are based on the following assumptions outlined in an unpublished memorandum dated December 8, 1980, from the Administrators of the National Highway Traffic Safety Administration and the Federal Highway Administration to the Regional Administrators of both agencies:

1. Speedometer error—the difference between the actual speed and the speed shown on the speedometer—ranges from ±5 mph,

2. There are an equal number of positive and negative errors,
3. The distribution of errors is uniform rather than unimodal, and
4. The distribution of vehicles traveling between 55 and 60 mph is uniform.

The adjustments only apply to the speed data for the percent of traffic traveling between 55 and 60 mph. This eliminates from those counted as violators any drivers who unknowingly exceed 55 mph because of an inaccurate speedometer. As stated in the memorandum outlining the adjustments, "positive speedometer error (speedometers that show a higher speed than the actual) need not be adjusted because vehicles exceeding 55 mph will indicate speedometer speeds higher than actual travel speeds." Thus, a driver who knowingly exceeds the speed limit should be included in the "percent exceeding statistic calculation." The apparent intent of adjusting only for negative speedometer error is to give the states the benefit of the doubt and to lower the percent exceeding to account for those who unknowingly break the law.

No adjustment is made for the opposite speedometer error. According to the assumptions, there will be drivers whose actual speed is less than 55 mph who, because of speedometer error, think they are traveling between 55 and 60 mph. To illustrate, consider the possible sources of error given in Table 47. The adjustment accounts for group 2, because they are unintentionally in violation of the law. Drivers in group 4 are clearly violators, both because they think they are speeding and because they are actually speeding. Group 3 is "knowingly exceeding the speed limit," although in actuality they are in compliance. According to the assumption that the speedometer error ranges from ±5 mph in equal proportion, the size of group 3 would cancel group 2. Thus, if group 3 were included there would be no reason for an adjustment for speedometer error. This same conclusion can be demonstrated mathematically (Appendix E).

TABLE 47 Possible Sources of Speedometer Error

Speed Shown	Actual Speed	
	Legal	Illegal
55 mph or less	1. Obeying law	2. Unintentional violation
55 mph or more	3. Obeying law, perhaps unintentionally	4. Intentional violation

The assumptions that underpin the adjustments do not appear fully warranted. The only study of speedometer error for vehicles in use revealed that speedometer errors range from ±10 mph, although the majority (90 percent) falls within ±5 mph (*23*). Thus the range of errors assumed in the adjustments is defensible. But the assumption of a uniform distribution of errors appears doubtful, because the error in speedometers revealed in the Auto Club study approaches a bell-shaped distribution (Figure 26).

Sampling Error

The states follow a required sampling plan in developing their estimate of state compliance. Monitoring stations located on each type of highway measure vehicle speeds on one day each quarter. Because states sample a relatively small proportion of total highway traffic, a degree of error exists when these data are generalized to the entire state. The range of possible error in sample data is routinely expressed as a confidence

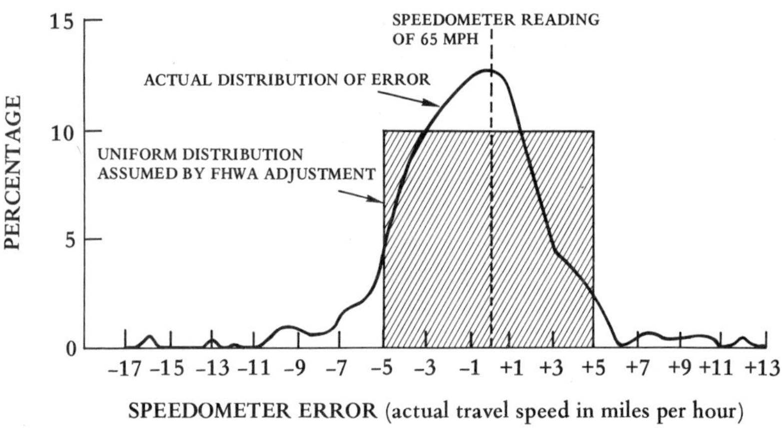

NOTE: The speedometer error shown was derived from four random samples of 800 vehicles in use in 1973. Speedometer error was tested at 25, 45, and 65 mph, and the variance increased with the speed tested. Hence, the variance at 55 mph would be somewhat smaller than indicated here. The distribution shown is based on four random samples of 200 vehicles each representing vehicles manufactured by GM, Chrysler, Ford, and American Motors. The sample has been weighted by the percentage of sales of each manufacturer in 1982 to more accurately reflect the fleet of American-made vehicles in use.

SOURCE: Automobile Club of Southern California, "Speedometer and Odometer Accuracy Investigation," 1974.

FIGURE 26 Distribution of speedometer error in American-made passenger cars compared with uniform distribution in speedometer error assumed by FHWA (*23*).

interval based on the sample standard deviation and the size of the sample. Generally a two-tailed test is used because the error could be either positive or negative.

Because Congress specifically requires a penalty against states that fail to meet the percent-exceeding standard, every effort should be made to avoid penalizing a state because of the error inherent in the data. For this reason the states are allowed to calculate the maximum overstatement of error in the data and reduce the derived estimate of statewide percent exceeding by that amount. To illustrate this adjustment, assume a state calculates that 52.6 percent of its drivers were speeding in any given year. With 40 sample sites, and a sampling error of 3.2 percent, it would have a downward adjustment of 5.37 percent to account for sampling error. This assumes a one-tailed test and a 95 percent level of confidence. This downward adjustment would lower the state's measure of compliance from 52.6 to 47.5 percent.

This adjustment appears reasonable given the intent of the adjustment. A state with a percent-exceeding measure of 49 percent might have a confidence interval of ±3 percent. Because the actual value could fall anywhere in that range (as high as 52 percent), the federal government would be hard pressed to justify exacting a penalty. The same applies to a state with a statewide percent-exceeding of 51 percent and a range of error of ±3 percent. The value might be as low as 49 percent or as high as 54 percent.

Granted the validity of the adjustment, it nonetheless introduces a counterincentive to accurate speed measurement. Some states have a standard error of about 3 percent, whereas others have errors of 6 percent or more, with the size of the error determined by the number of sampling stations. Each state was originally required to increase the number of sample sites to the point that its standard error would diminish to ±2 percent. After reviewing the state sampling plans for 1982, however, the FHWA decided that its target precision was not obtainable without dramatically increased expenditures by some states. It then rescinded the requirement that each state have a sufficient number of sample sites to meet the target level of precision and froze the number of stations at the existing level. It justified the decision by stating that the costs would exceed the benefits (*23*).

An example of the variation in precision can be seen in Texas, which has about 100 speed-monitoring stations and a sampling error adjustment of 3.12 percentage points. Meanwhile, Virginia, with about 36 speed-monitoring stations, has an adjustment of 7.7 percentage points. Thus, Virginia would be considered in compliance (based on sampling error alone) if unadjusted speed data showed 57.7 percent of traffic exceeding 55 mph, whereas Texas would not be in compliance if 53.3 percent of traffic exceeded 55 mph. Obviously states such as Virginia, which have fewer sampling stations, have an advantage over other states because of their larger adjustment.

Machine Measurement Error

Automated loop monitors may have a small degree of error. To date no state has successfully argued that such an error biases the measure of vehicle speed. After extensive testing of loop monitors, the FHWA contends that such error, if it exists, would be very small and randomly distributed.

Potential Sources of Distortion

Speeds are monitored in many different locations and the resulting measurements must be combined to compute the statewide average. In computing this average, the individual data are weighted based on the volume of traffic on the individual highway systems being monitored. The resulting averages could be rendered inaccurate if incorrect weights are applied to the speed data collected on each type of highway. The weights are derived from the amount of vehicle miles traveled (VMT). The FHWA does not monitor the development of the VMT estimate as carefully as it does the speed data, but there is sufficient disincentive to lowering VMT estimates, because federal aid for highways is partially apportioned on this basis. Nevertheless, details of the weighting process could be altered to reduce the measured noncompliance.

In 1982, for example, the state of Nevada argued that because of an expansion of its urban boundaries by the Bureau of the Census, some of its highways that had been classified as rural should be reclassified as urban. These formerly rural highways and their VMT increased the weight assigned to the urban speed data. Because urban speeds tend to be lower than rural speeds, by entering this new estimate into the calculatons, Nevada's adjusted estimate of noncompliance declined from 51.2 to 50.0 percent.

Adjustments for Tolerance

Although the current adjustments overstate the error caused by inaccurate speedometers, they do not account for the tolerances traditionally allowed in speed-limit enforcement. These tolerances are deeply ingrained into the deterrence mechanism because of the practices of traffic courts and the police as well as the expectations of the public.

Under the current standard an individual traveling at 59 mph is counted with the same weight as one speeding along at 75 mph. This is inappropriate for two reasons: (a) the first driver is within the tolerance zone allowed by almost all traffic police—a zone that would exist at any speed limit, and (b) the safety risk is much greater for the driver traveling 75 mph. The current overstatement of speedometer error in the compliance data accidently adjusts for tolerance, but it would be preferable to be explicit about adjusting for tolerance in a straightforward manner.

Summary of Current Adjustments

The adjusted data do not accurately reflect actual vehicle speeds on the highway, nor are they intended to do so. Instead, the data reflect a measure of willful noncompliance, with generous allowance for sampling error. The FHWA requires that the states report both adjusted and unadjusted speed data. When the agency publishes statistics on speed trends on the nation's highways, it uses the unadjusted figures. The adjustments can be defended as a reflection of congressional intent to avoid penalizing states for measurement errors.

The adjustment for sampling error illustrates the agency's erring on the side of caution to avoid unfairly penalizing states because of the statistical properties of sample data. The adjustment for speedometer error, on the other hand, appears flawed and should be dropped. Overall the adjustments appear to have become a mechanism for ensuring that the schedule of compliance, which is set in federal law, is attainable. This may reflect a compensation for the lack of explicit allowance for enforcement tolerances in the federal standards.

Organizational Mismatch

In addition to technical shortcomings in the mechanism by which the federal government measures state compliance with the 55 mph speed limit, there is also an organizational mismatch between the responsibility for compliance and the incidence of penalties for noncompliance. In particular, state highway patrols and other police agencies are responsible for enforcement of the law, whereas state departments of transportation are threatened by a loss of funds if a state is found in noncompliance. Under current law a state can lose 10 percent of its urban, primary, and secondary funding (about 3 to 4 percent of its total highway aid) if the standard is not met in any single year. This creates awkward organizational pressures. Even though the threat of lost highway funds may encourage the governors to require adequate enforcement by the state patrols, the agency that has responsibility for implementing the policy is not the one that would suffer if motorists fail to comply with the speed limit. Agency budgets and staffing plans may severely restrict the ability of highway patrols to realign their enforcement priorities. To help ease such pressures, the National Highway Traffic Safety Administration encourages enhanced enforcement by awarding grants for this purpose to state patrol agencies. This program appears to be too small to be of much consequence in compensating for the additional costs involved in enforcement of the 55 mph speed limit, however. In fiscal year 1982 obligations for enforcement of the 55 mph speed limit totaled only $17 million, or only about 4 percent of the states' estimated cost of enforcing the speed limit.

Second, both state and local police are responsible for enforcement of the speed limit, whereas federal penalties for noncompliance are the responsibility of the states. In many states enforcement of speed limits on urban highways and rural two-lane routes is primarily the responsibility of the local government. These highways reported the greatest increase in speeding in 1983, and the agencies primarily responsible for enforcement of the speed limit on these highways are not covered by federal regulations. The role of the traffic judges is also independent of the federal role.

In brief, federal penalties for noncompliance are poorly linked to responsibility. They are linked to only one level of government (states) and only one area of jurisdiction (highway programs), even though much of the process for inducing compliance lies outside the organizational range embraced by federal penalities.

Misallocation of Enforcement Resources

The current compliance standards encourage states to focus their enforcement efforts on Interstate highways, which, in turn, draws resources away from less safe secondary roads. From the standpoint of demonstrating compliance, this is an appropriate response: in many states the Interstates carry more than one-half of all travel on highways posted at 55 mph, even though they make up less than 5 percent of all highway mileage posted at 55 mph. By focusing resources on the Interstates, enforcement agencies can influence more drivers per patrol officer deployed.

Although this is an efficient allocation for gaining compliance, from the standpoint of overall highway safety, it represents less than an optimal allocation of resources. Such allocation may imply diminished enforcement of speed limits on highways where speeding may be a more serious safety problem. For example, the Oregon state police have indicated that they devote one-third of patrol time for enforcement of the 55 mph speed limit on freeways, even though only 6 percent of fatal accidents occur on the freeways (*24*).

Interstate speed-limit enforcement does not necessarily drain resources from other activities. Some states rely on supplemental funds for enforcement of the 55 mph speed limit; these funds are used to pay overtime to patrol officers. However, state patrol agencies report that only a handful of states rely on overtime enforcement, and even in these states the overtime amounts to at most 10 to 15 percent of patrol manhours and usually less than 5 percent (Appendix F).

Some states have reallocated personnel from patrols on other systems. The International Association of Chiefs of Police reported that one-half of the states in its survey redeployed police resources for speed-limit enforcement (*25*). Officers in various states made the following comments on the pressures that enforcement of the 55 mph speed limit places on other activities:

> Approximately 60 percent of Delaware State Police enforcement effort was directed at 3.6 percent of Delaware's highways during the National Maximum Speed Limit project. During the first 8 months of 1980, there was a 24 percent increase in fatal accidents on non-National Maximum Speed Limit highways.
>
> [North Carolina] Fifty percent of traffic collisions are occurring on the secondary highway system, but the National Maximum Speed Limit forces 75 percent of man hours [to] be directed to the primary system.
>
> [Texas] It has directed traffic enforcement from rural highways to a large percentage of time devoted to Interstates that have only half of the fatal accident frequency problem as rural roads.
>
> [Washington] Demands in all categories have increased since 1972; however, because of budgetary constraints we now have [fewer] troopers to confront inceasing problems with the 55 mph National Maximum Speed Limit. With its compliance mandates, the National Maximum Speed Limit consumes departmental attention out of proportion with the safety related significance of speeds between 55 and 65 mph on open highways (*25*).

Many states do not find that the 55 mph speed limit draws resources from other police responsibilities. The officers in these states, however, point out that they do not make concentrated enforcement efforts to maintain compliance. Although enforcement of the 55 mph speed limit has not resulted in a misallocation of safety-related enforcement in all states, it has caused a shift in enforcement priorities in others, perhaps to the detriment of overall highway safety. This problem could be substantially alleviated if the federal measure of state compliance was more reflective of the safety implications of individual violations, as discussed in the following section.

Alternate Compliance Measurement System

The existing compliance standard emerged from the schedule for compliance outlined in the Highway Safety Act of 1978. The Congress expected that after a few years of effort the states would be able to hold 70 percent of drivers at or below 55 mph. The standard was later lowered to 50 percent when it became apparent that states could not achieve the schedule originally set forth. Today even the 50 percent standard appears to be unachievable in some states, as reflected in the growing dependence on adjustments to achieve apparent compliance.

Because speed variance, as much as speed itself, influences highway safety, much of the safety benefit of a speed limit is derived from the more uniform pace of travel induced by the limit. It may be relatively unimportant whether the average speed is a little higher or a little lower as long as all the traffic moves along at the same speed. Thus, the existing standard would be improved if it encouraged a narrower variance in traffic speeds while making allowances for enforcement tolerance.

One approach to handling this problem is to weight the number of drivers traveling at high speeds more heavily than those drivers who are barely traveling above the speed limit. As an illustration of this concept, a state would be found out of compliance if it scored more than a set number of points (300, for example) using the following point system:

Percentage of Motorists Traveling:	*Weight By*
55 to 59 mph	1 point
60 to 64 mph	5 points
65 mph or more	10 points

Under such a system, the percentage of motorists exceeding the speed limit would continue to be the basis for the adjustment, but increased stress would be placed on the faster driving that increases the variance in speeds and most impairs safety. Under this approach, a state in which 100 percent of motorists traveled at 57 mph would score 100 points (1 point × 100 percent). This state would not be in compliance under the existing regulations, even though having all motorists traveling at a uniform speed would improve highway safety. Conversely, if two-thirds of a state's motorists travel at speeds below 55 mph but the remaining one-third travel at 65 mph, this state would be in compliance under current regulations but would fail the proposed point system.

Yet this speed distribution appears to be less safe because of the wider variance in speeds. A point system such as the preceding one could thus help to build a sanctions mechanism that more closely reflects safety risks and safety priorities. The total number of allowable points could be set to reflect sampling error or any other necessary adjustment. Extensive analysis will be required to fine-tune this concept, but in principle it appears to be a much better approach to measuring compliance.

Further, such a point system could also be used to reflect the tolerances that traditionally have been granted by the police and the courts. Citations issued for speeding at less than 5 mph above the posted speed limit are awkward for the police, may not be upheld in traffic courts, and are irritating to motorists. Unfortunately the current federal sanctions process may encourage just this type of unrealistic behavior. To a large extent, the incentive to cite incremental violations has been dampened by the liberal adjustments process, which makes allowances that, in effect, compensate for the failure of the standards to reflect tolerances explicitly. A more direct and even-handed way to reflect the tolerances that the public, the police, and the courts have accepted would be to build them directly into the sanctions process. For example, a small weight—even zero—could be applied to speeding violations that fall within the permissible tolerance.

In addition to reflecting tolerances more realistically, and treating violations in closer relation to their safety risks, a point system could be extended to reflect safety priorities on different types of highways as well. Fatality rates on two-lane highways are from two to four times greater than fatality rates on Interstates, and speeding on these highways should be emphasized in enforcement. By weighting the high-speed drivers on these highways more heavily, the states would be encouraged to devote their enforcement resources where the safety risk is the greatest.

CONCLUSION

Part of the impetus for this study came from congressional concern about states that enacted low fines for offenders of the 55 mph speed limit. Whether such practices have an effect on speeds and safety remains uncertain. On the basis of the level of fines authorized by state law and on whether points are assessed for speeding and entered into the offending driver's record, states were classified as high-penalty, medium-penalty, and low-penalty states. The low-penalty states as a group had higher speeds and higher fatality rates than the other two categories. But some low-penalty states had lower speeds and lower fatality rates than many of the states in the other groups. Further, weak penalties were not necessarily the cause of low compliance: the average percent speeding on Interstate routes varies only randomly with respect to state statutory penalties.

Many other factors—actual police practices, court behavior, public awareness, and the like—influence how laws affect driving behavior. Additionally, state laws and driver behavior may both reflect some third factor, such as public attitudes. Even so, many of the low-penalty states issue more citations for speeding per vehicle mile of travel than other states, which suggests that they are not shirking their responsibility for enforcement.

The committee concludes that there is no identifiable relationship between state statutes and compliance with the 55 mph speed limit. Instead, this link is immersed in a complex, interconnected set of relationships among laws, enforcement policies, court practices, and public attitudes. The committee also takes note of past experience with sharply increased penalties and enforcement. Such attempts to alter driver behavior can have unexpected and adverse consequences. Experience with crackdowns on drunken driving and speeding in the past has indicated that the police and the courts are unwilling to apply penalties they perceive as overly severe and out of proportion to the crime.

Of all the actions that can be implemented to induce compliance with the speed limit, increasing the visibility of enforcement appears most likely to be effective. Because there is only one on-duty state highway patrol officer for every 190 miles of highway posted at 55 mph, there are clearly practical limits to how ubiquitous enforcement can be. Nevertheless, experimentation with different patroling strategies indicates that visible enforcement does yield a difference in compliance. The implication for federal policy is that federal attention is more appropriately focused on the saliency of enforcement rather than the severity of state laws. The current levels of popular support are conditioned on existing levels of enforcement. It is not clear whether the majority of the public would continue to support the speed limit if enforcement increased dramatically.

At present, federal law requires the withholding of federal highway funds from a state in which more than 50 percent of all motorists on roads posted at 55 mph exceed the speed limit. Although various adjustments permitted under current law make this requirement less stringent than it first appears, this method of scorekeeping is fundamentally out of tune with traffic court practices, traditional enforcement tolerances, and public expectations. In spite of the different shadings of importance that individuals and organizations attach to speeding offenses of different severity, the current federal sanctions mechanism makes no explicit allowance for this tradition.

The current mechanism also fails to reflect safety priorities. The risk to safety is much greater for someone driving at 80 mph than it is for someone driving at 56 mph, but current federal procedures count both violations equally. To reflect both safety priorities and traditional organizational practices, the committee recommends that the federal government measure state compliance with the speed limit through a point system that attaches more significance to high-speed violations than to violations just above the speed limit. Such a point system could also be extended to reflect the different safety risks on various road systems.

The existing federal mechanism for monitoring state compliance with the 55 mph speed limit, by which states that experience poor observance of the 55 mph limit are threatened by federal sanctions, is increasingly leading to an untenable situation. In 1983, of the states that reported having a majority of traffic on roads posted at 55 mph travel in excess of 55 mph, all 37 were able to adjust their data to appear to be in compliance. Without the spurious elements in the adjustments, many states would not be in compliance. It appears that liberal adjustments of dubious statistical merit have been used to avoid situations where the federal government would be forced to penalize the states by withholding funds.

Adjustments for speed-limit enforcement tolerance should replace the existing adjustment based on speedometer error. Such a change would obviate the need to adjust for inaccurate speedometers. In addition, the current adjustment for sampling error, although valid, is not applied fairly in all states. Some states have a wider adjustment because they do not have large enough samples to reduce their sampling error to a level achieved by others. The standard for sampling error should be set at some level that all states can achieve, and should result in an equal-sized adjustment.

A more risk-related weighing scheme of the type proposed in this chapter would encourage law enforcement agencies to allocate their resources so as to reduce the points consumed against them. This means that proportionally more resources would be directed at high-speed offenders and off-Interstate offenders. Such a reallocation appears desirable in many instances and would redress one of the charges now made against the 55 mph speed limit: namely, that it encourages police agencies to misallocate resources to offenses that present relatively minor safety risks. In addition, by adopting the weighed formula approach, the widely practiced tolerance in speed-limit enforcement will be built into the compliance mechanism, and states will be encouraged to maintain a narrower variance in speeds on their highways.

REFERENCES

1. *Highway Statistics*. Tables VS-1 and VS-2 and Appendix, FY 1983 Speed Monitoring Data Summaries. FHWA, U.S. Department of Transportation, 1980–1982.
2. H.L. Ross. "Traffic Law Violations: A Fold Crime." *Social Problems*. Winter 1960–61, Vol. 8, No. 3, 1961.
3. J.Q. Wilson. *Thinking About Crime*. Rev. ed., Basic Books, New York, 1983.
4. D.T. Campbell and H.L. Ross. "The Connecticut Crackdown on Speeding Time Series Data in Quasi-Experimental Analysis." *Law and Social Review*, No. 3, 1968.
5. *55 mph Fact Book*. NHSTA, U.S. Department of Transportation, 1979.
6. Field Research Corporation. *A Study of California Public Opinion Concerning Automobile Safety Belts and the 55 mph Speed Limit*. San Francisco, 1977.
7. M. Edwards and R. Brackett. *Comparative Evaluation of Speed Control Strategies*. Texas A&M University, College Station, 1977.
8. California Highway Patrol. *Major Corridor Overtime Speed Enforcement Project*. Final Report. Planning and Analysis Division, Long Range Planning, Caltrans, Sacramento, 1982.
9. G.W. Duff. New York State Police Controlled Access Highway Task Force. *Second Symposium on Traffic Safety Effectiveness (Impact) Evaluation Projects*, Fredericksburg, Va., 1982.
10. T.M. Klein. *An Evaluation of the 55 mph Speed Limit Enforcement Demonstration Projects in Connecticut and Utah*. HS-806-151. NHTSA, U.S. Department of Transportation, 1981.
11. S. Baker. The Effect of Enforcement on Vehicle Speed. *Bull. 91*. HRB, National Research Council, Washington, D.C., 1954.
12. R.D. Smith. *The Effects of Enforcement on Driving Behavior*. International Association of Chiefs of Police, Washington, D.C., 1962.
13. A. Calica, R. Crowther, and K. Schumate. *Enforcement Effect on Traffic Accident Generation*. State Department of Public Administration, Bloomington, Ind., 1963.

14. F. Council. *A Study of the Immediate Effect of Enforcement on Vehicular Speeds*. Highway Safety Research Center, University of North Carolina, Chapel Hill, 1970.
15. F. Council. *Police Traffic Services*. Highway Safety Research Center, University of North Carolina, Chapel Hill, 1971.
16. T. Joscelyn and D. Goldenbaum. *A Study of the Effects of Law Enforcement on Traffic Flow Behavior*. Institute for Research in Public Safety, Bloomington, Ind., 1971.
17. B.J. Campbell. *Speed Compliance During Operation CARE*. Highway Safety Research Center, University of North Carolina, Chapel Hill, 1979.
18. N. Darwick. *Enforcement of the National Maximum Speed Limit: Practices and Procedures*. International Association of Chiefs of Police, Gaithersburg, Md., 1977.
19. Teknekron Research, Inc. Compendium of Survey Items and Results of Safety Countermeasures. TK 76-1045R. NHTSA, U.S. Department of Transportation, 1979.
20. Title 23, United States Code, Section 154(f), as amended in the Highway Safety Act of 1978, P.L. 95-599, Sec. 205(e).
21. Omnibus Reconciliation Act of 1981, Section 1108(b). (Pub. L. 97-35).
22. Title 23, United States Code. Section 154(e).
23. Automobile Club of Southern California. *Speedometer and Odometer Accuracy Investigation*. Public Safety Department, March 21, 1974.
24. *Federal Register*, Vol. 47, No. 214, Nov. 4, 1982.
25. R. Sostkowski, D. Phillips, R. Monroe, and A. Pearson. *Determining the Cost of Enforcing the National Maximum Speed Limit*. Appendix C. International Association of Chiefs of Police, Gaithersburg, Md., 1980.

10
Recommendations and Conclusions

The 55 mph national maximum speed limit, enacted during the Arab oil embargo of 1973 to conserve fuel, became effective in all states by March 1974. Motorists slowed down and the number of highway fatalities declined from 55,511 to 46,402 per year—an unprecedented decrease in highway fatalities outside of wartime. Because of these unexpected safety gains, Congress made the 55 mph maximum speed limit a permanent pillar of the nation's highway safety policy.

Although the 55 mph speed limit contributed to the safety improvements of 1974, it does not account for all 9,100 fewer fatalities in that year. The complexity of accidents, limitations of aggregate statistics, disruptions in travel patterns that accompanied the oil embargo, and the confounding effect of the trend in safety, preclude determining a single specific number by which the 55 mph speed limit reduced highway fatalities when first introduced. Nonetheless, estimates from the many studies reviewed in Chapter 3 suggest that from 3,000 to 5,000 lives were saved by the speed limit in 1974. The current estimate of the impact of the speed limit on safety places the number of lives saved closer to 2,000 to 4,000.

The impact of the speed limit on serious injuries is somewhat more difficult to determine because of limited trend data and inconsistencies in injury definitions. The best guess of the current effect of the speed limit is that about 2,500 to 4,500 fewer serious, severe, and critical injuries and 34,000 to 61,000 fewer minor and moderate injuries occur each year because of slower and more uniform speeds. As a result of fewer injuries, medical costs are reduced by some $50 to $90 million annually (Chapter 4).

In addition, as a result of the 55 mph speed limit about 167,000 barrels of petroleum are saved per day. This reduces petroleum consumption in the transportation sector by 1.8 percent annually and reduces national consumption of petroleum by about 1 percent (Chapter 6).

The 55 mph speed limit also reduces the cost to taxpayers of several government programs. Medicare, Medicaid, and Old Age Survivors and Disability Insurance have reduced costs of some $52 million annually due to fewer motor vehicle fatalities, disabilities, and injuries. When benefits to other governmental budgets at the national, state, and local levels are included, the total savings to taxpayers are about $65 million each year (Chapter 5).

The many benefits to society of the 55 mph speed limit are accompanied by considerable costs as well. The additional travel time caused by slower travel in 1982 amounts to about 1 billion hours compared to 1973 speeds. Each driver in the United States spends, on average, about 7 additional hours per year on the road because of slower speeds (Chapter 7).

In summary, the speed limit reduces total highway fatalities by 4 to 8 percent, reduces petroleum consumption in the transportation sector by 1.8 percent, reduces medical accident costs by 1 to 2 percent, and reduces the costs of motor vehicle accidents borne by the taxpayers by 3 percent. The cost of these benefits is a 1.3 percent increase in motorist travel times (Table 48).

RECOMMENDATIONS

The safety benefits of the 55 mph speed limit led the Committee for the Study of Benefits and Costs of the 55 mph National Maximum Speed Limit to recommend that the law be retained. The range of estimated fatality reductions indicates that some 25,000 to 50,000 lives have been saved as a result of the speed limit during the last 10 years. Few safety policies can rival the impact of the 55 mph speed limit in reducing the accidental deaths of Americans as they travel about the country.

The committee unequivocally endorses the law as it applies to urban highways and rural arterials and collectors. These highways make up 94 percent of the mileage posted at 55 mph. The committee debated at great length whether its support for the 55 mph speed limit should include all of the nation's highways or whether it should exclude certain

TABLE 48 Current Impact of the 55 mph Speed Limit (*1*, *2*)

Major Impact Areas	Total (1983)	Reduction due to 55 mph Speed Limit (1983)	Percent Reduction
Highway fatalities	44,300	2,000–4,000	4–8
Transportation petroleum consumption (000s of barrels per day)	9,200	167	1.8
Medical costs of motor vehicle accidents ($ million)	4,440	50–90	1–2
Property damage and legal costs ($ million)	30,000	72–150	0.3–0.5
Motor vehicle accident costs borne by taxpayers ($ million)	2,500	65	3
		Increase Due to 55 mph Speed Limit 1983	Percent Increase
Travel time (billions of passenger hours)	75.0	1.04	1.3

NOTE: Estimate of 1983 fatalities provided by the National Safety Council.

roads that are relatively lightly used and designed for high-speed driving. The committee recognized that the trade-off of the benefits of the speed limit and its costs are not equal for all highway systems. In addition, acceptance of the law varies among certain groups and across regions of the country.

Safety Impact and Travel on Various Highway Systems Posted at 55 mph

The chief objection to the 55 mph speed limit is that it sets the same standard for the highest quality Interstate highway as for a narrow, two-lane rural road. This blanket standard applies to highways that are vastly different in terms of the relative risks to safety and their importance to national travel. For example, rural Interstates, which carry 19 percent of travel on highways posted at 55 mph, and which have the highest average speeds, only account for 9 percent of fatalities on roads posted at 55 mph. In contrast, rural arterials, which carry 30 percent of travel, account for 38 percent of fatalities (see Table 49). Much of the discrepancy in safety can be explained by the differences in the physical standards of these highway types. Rural Interstates have been built to the highest standards, and feature multiple lanes, wide medians, and controlled

TABLE 49 Total Travel and Fatalities on Highways Posted at 55 mph (1982)

Highway System	Urban: Vehicle Miles of Travel (millions)	Urban: Total Fatalities	Rural: Vehicle Miles of Travel (millions)	Rural: Total Fatalities
Interstates	157,974	1,644	140,778	1,983
Urban freeways[a]	73,998	740	—[a]	—[a]
Arterials[b]	42,243	1,335	223,903	8,298
Collectors/ Local[b]	—[c]	273	106,950[b]	7,385
Total	274,215	3,992	471,631	17,666

NOTE: Estimates of 1982 fatalities classified by highway systems posted at 55 mph were provided by the National Center for Statistics and Analysis from the Fatal Accident Reporting System, National Highway Traffic Safety Administration. Estimates of travel by highway posted at 55 mph were provided by the Federal Highway Administration, Office of Highway Planning from the Highwy Performance Monitoring System (HPMS).

[a]Classification applies to urban freeways only.
[b]Generally two-lane roads.
[c]Travel estimated for collector routes only. Estimates of travel on local routes posted at 55 mph are not available.

access. Arterial routes tend to be two-lane roadways with at-grade intersections. By setting the same maximum limit for all highways without regard for their geometric design and safety performance, the speed limit reflects an uneven commitment to safety across road systems.

Support for and Opposition to the National Maximum Speed Limit

Although many motorists express dissatisfaction with the slower speeds and increased travel time caused by the 55 mph speed limit, the majority of the public supports the law. The most recent national poll reveals that 76 percent of Americans endorse the 55 mph speed limit.

Although more than two-thirds of the public favors the law in every region, the support is strongest in the Northeast and weakest in the South and West (*3*).

Region	*Favor*	*Oppose*	*No Opinion*
Northeast	81	17	2
Midwest	75	24	1
South	70	28	2
West	72	25	3

Residents in rural areas tend to depend more on automobile travel and tend to spend more of their time driving on highways posted at 55 mph. Motorists in the Mountain West region of the country spend 15 percent of their travel time on rural Interstates as compared to only 6 percent by motorists in the Northeast. The somewhat greater opposition to the 55 mph speed limit that is evident in the Southwest and West may reflect the additional time costs borne by drivers in these states relative to drivers in the Northeast.

Public support for the law diminishes with the amount of driving done each year. Whereas more than 80 percent of those who drive less than 5,000 miles a year support the law, only 54 percent of those who drive more than 30,000 miles each year support it (Table 50). Traveling salesmen, independent truckers, regional managers, and others who depend on Interstate travel for their livelihood bear a disproportionate time burden because of the 55 mph speed limit.

Although the committee endorses the 55 mph speed limit on most of the highways now posted at 55 mph, the members were divided on whether the speed limit should remain intact on all roadways. Some judged that adjustments should be permitted on rural Interstate highways to ameliorate the frustrations and the costs of the law. Among these, some believed that such adjustments should be made only when other compensating safety policies were enacted. The debate about whether exceptions should be permitted on rural Interstates focused on two general questions. First, is a change in policy being rendered necessary by the trend of increased driving speeds, reduced compliance as measured by automatic speed monitoring devices, and growing state resistance to effective speed-limit enforcement? Second, are changes in national policy desirable because of uneven trade-offs on various high-

TABLE 50 Attitude Toward the 55 mph Speed Limit by Amount of Driving Each Year (*4*)

Miles Driven	Strongly Favor (%)	Somewhat Favor (%)	Somewhat Oppose (%)	Strongly Oppose (%)
0–2,000	74.4	14.4	5.6	4.2
2,000–5,000	60.5	21.9	11.2	5.6
5,000–10,000	60.5	19.8	10.5	8.3
10,000–15,000	56.2	21.3	12.6	9.9
15,000–20,000	52.0	26.7	10.0	11.3
20,000–30,000	47.8	20.9	14.2	17.2
More than 30,000	36.1	17.6	23.1	20.4

way systems, and public attitudes favoring return of speed-limit jurisdiction entirely to the state and local governments?

IS CHANGE NECESSARY?

Whether the current laws governing the 55 mph speed limit need to be altered involves three questions:

- Are the benefits of the 55 mph speed limit being eroded by higher speeds and other influences on safety?
- Will eroding compliance and motorist frustration lead to the ultimate nullification of the law?
- Does the widespread violation of the 55 mph speed limit breed disrespect for law?

Are the Benefits of the 55 mph Speed Limit Being Eroded by Higher Speeds and Other Influences on Safety?

The safety benefits of the 55 mph speed limit were estimated early in its existence. Average speeds have since increased, and many improvements to vehicles, roadways, and medical services have taken place. Three factors influence the size of the benefits of the 55 mph speed limit: (a) the behavior of motorists as measured by average speeds and speed variance, (b) other improvements in vehicle crashworthiness and roadside improvements that minimize crash severity, and (c) increased exposure to risk due to increased travel on highways posted at 55 mph.

Motorist Behavior

Since 1974 the average highway speed has been increasing. Considering conditions before enactment of the 55 mph speed limit, recent trends indicate that about 60 percent of the reduction in average speeds that was achieved in 1974 has been retained.

The sketchy information available on the trend in speed variance suggests that it may have remained relatively constant, in effect shifting the speed distribution without increasing the dispersion. Trend data on speed variance are insufficient to document this point for all highway systems. Reclassification of the federal-aid highway system in 1976 and changes in reporting formats have interrupted the trends for all but the Interstate highways. Trend data are available for rural and urban Interstate highways, but many changes have occurred in speed measurement techniques, the completeness of reporting by individual states, and in the sampling methods from which estimates are drawn. Nonetheless, given the sharp discontinuity from past trends (discussed in Chapter 2), it appears that the shift in speed variance that occurred in 1974 has been sustained.

There is little statistical basis for estimating the effect on safety of the increases in average speeds, particularly because correspondingly large increases in variance have not occurred. It is probably fair to assume that, at most, the erosion in safety benefits is proportional to the increases in average speed relative to the reductions obtained in 1974. This suggests that the effectiveness of the speed limit may have diminished by 40 percent during the last decade. If instead, the safety benefits are proportional to the variance of speed, then the speed limit may not have lost any of its effect on safety. Therefore, depending on what is assumed about the relationship between driver behavior trends and speed, the speed limit could be anywhere from 0 to 40 percent less effective today than in 1974. In the absence of better information, a 20 percent reduction in effectiveness provides a reasonable approximation of the reduced effectiveness of the 55 mph speed limit because of increased speeds.

Other Improvements in Safety

During the last decade the fatality rate (deaths per 100 million vehicle miles) for all roads and highways has declined by 23 percent, from 3.59 in 1974 to 2.76 in 1983. Much of this improvement has been caused by continuing improvements to vehicles, roadways, and medical services. These safety improvements suggest that crashes at higher speeds may be less severe today than 10 years ago, which, in effect, reduces the importance of the 55 mph speed limit. If the safety benefits associated with the reduction in the speed limit declined in proportion to the overall fatality rate during the last decade, then the 55 mph speed limit would be perhaps 23 percent less effective in reducing current fatality rates than in 1974.

Increased Travel

Travel has increased approximately 24 percent during the last decade. With an increase in the number of motorists on the road, the number of people exposed to risk has increased. As a result, more people are being protected by the reduced speeds of the 55 mph speed limit.

Multiplying the combined effects of these changes—a 20 percent reduction in the effect of the speed limit due to speed changes, a 23 percent reduction because of other safety policies, and a 24 percent increase in the number of motorists exposed to risk—suggests that the speed limit saves 2,000 to 4,000 lives annually. Although these mechanical adjustments oversimplify the complexity of the safety trends of the last decade, they illustrate that in approximate terms the benefits of the speed limit remain large. The force for changing the law may well come from other pressures, such as motorist frustration and federal-state conflict concerning sanctions, than from diminishing safety benefits attributable to the 55 mph speed limit.

Will Growing Noncompliance Lead to Nullification of This Law?

One of the greatest pressures to increase speed limits is the behavior of the majority of motorists. Nearly three-fourths of drivers drive at speeds in excess of 55 mph on rural Interstates. The percentage of motorists speeding on state highways increases every year. The states' ability to enforce this law is limited: with 190 miles of highway for every on-duty state patrolman, large segments of the highway system are of necessity lightly patrolled. Compliance with the law on most highways depends on the voluntary actions of individual motorists who are apparently increasingly unwilling to obey the speed limit.

These trends in speeding will ultimately result in some states failing to meet the compliance standard, even given the full benefit of current (dubiously valid) adjustment procedures. In 1981 Nevada had to make numerous special adjustments to reduce its estimate of speeding to 50 percent (the mandated standard). In 1983 Michigan was allowed to use only a part of its sample of speed monitoring results to keep its figures within the technical limits. Oklahoma, initially out of compliance in its first 1983 certification report, was later able to lower its unadjusted percent exceeding statistic from 60.4 to 49.8 percent by using an alternate adjustment permitted by FHWA. All appearances indicate that ultimately some states will be ruled in noncompliance and threatened with loss of highway funds. This could increase the pressure to remove the federal oversight role in setting maximum speed limits. By making adjustments to speed limit policy, the nation may be able to hold onto the majority of the benefits of the law.

Supporters of the 55 mph speed limit argue that eroding compliance is not necessarily a harbinger of abandonment of the law, and that the use of the technical adjustments within current procedures indicates that the law is adjusting realistically to disparate regional demands. In addition, they argue that it is somewhat misleading to consider drivers within 5 mph of the limit as speeders. Public opinion polls reveal that most supporters of 55 mph believe they should be able to travel at 60 mph without being ticketed. In doing so they consider themselves in compliance with the spirit of the law. Whether the growing noncompliance threatens the integrity of the existing policy is a matter on which individual opinions differ. Yet the question of whether the status quo can be sustained lies at the heart of the controversy on whether to relax the 55 mph speed limit.

Does Increased Noncompliance with the 55 mph Speed Limit Breed Disrespect for Law Generally?

Those who believe that the 55 mph speed limit should be relaxed argue that preoccupation with this particular statute impedes the enforcement of more vital laws, confuses the public about the most important elements of highway safety, and encourages a misallocation of enforcement resources. Some argue that it is hypocritical to maintain the 55 mph speed limit while tolerating such widespread noncompliance, and that this could diminish public respect for other laws.

Advocates of retaining the 55 mph speed limit dispute that this statute is different from other statutes with respect to compliance and the zeal of enforcement. Speed limits, like all traffic regulations, are intended to regulate a safe and even flow of traffic. Unlike laws prohibiting behavior that is morally repugnant, such as burglary, rape, or murder, violation of traffic laws is not considered morally wrong by many. Drivers regularly speed without thinking of themselves as criminals. Rules of the road serve a clear, practical purpose by keeping drivers orderly. Most traffic laws, whether they require coming to a full stop at stop signs or slowing down in schools zones, are only loosely obeyed by the motoring public. For example, 80 percent of motorists fail to come to a full stop at stop signs at low-volume intersections, a statistic that improves to only 50 percent compliance at heavily traveled intersections (*5*). In spite of such statistics, few would argue that stop signs have no effect or that their continued use is threatened by widespread noncompliance. They do alter motorist behavior in desired ways (if imperfectly) and they are widely accepted by the public.

Summary

Scientific and technical analysis can only partly answer the question of whether the problems with the 55 mph speed limit are so great that the law should be modified. The answers to the key questions that underpin this discussion—whether the benefits of the law can be sustained, whether eroding compliance will lead to its nullification, and whether widespread violation of the law leads to disrespect of laws generally—involve forecasts and value judgments that are unavoidably controversial.

IS CHANGE DESIRABLE?

The committee also discussed at length the implications of changes to current policy to determine whether the benefits of such changes would make modifications to current policy desirable. This discussion involved three questions:

- What would be the consequences of higher speeds on rural Interstates?
- Could the increased speeds be contained to just those highways with higher speed limits?
- Should the federal government continue to set maximum speed limits?

What Would Be the Consequences of Higher Speeds on Rural Interstates?

Currently 31,500 miles of rural Interstate highway are posted at 55 mph; these routes represent about 6 percent of all mileage posted at 55 mph throughout the country and carry about 19 percent of all traffic on highways posted at 55 mph. Although average speeds are the highest on this system, the fatality rate, per mile traveled, is among the lowest. Rural Interstates have been built to the highest design standard of any class of highways. The direction of traffic is usually separated by wide medians or median barriers that greatly reduce the risk of head-on crashes. Rural Interstates have wide shoulders, breakaway sign posts, guardrails, and other physical safety features that minimize crash severity. Only urban Interstates have a lower fatality rate, in part because greater traffic density reduces speeds and speed variance. The number of highway deaths that occur on rural Interstates is small compared to the total number of deaths on all roads. Deaths on rural Interstates accounted for 9 percent of the fatalities on highways posted at 55 mph and accounted for about 4 percent of fatalities on all roads and highways (Table 51).

Estimates of the impact on safety of highway speeds on rural Interstates are necessarily somewhat rough. For purposes of illustration, the following assumptions have been made: (a) given the option to increase speed limits, all states would return to higher speeds on rural Interstates; (b) the increased fatality rates would be proportional to the decline in

TABLE 51 Motor Vehicle Fatalities on All Roads and on Highways Posted at 55 mph (1982)

Highway Type	Total Number	Percent of Total
All roads and highways	43,721	100.0
Highways posted at 55 mph	21,658	49.5
Non-Interstate freeways	740	1.7
Arterials	9,633	22.0
Collectors/local	7,658	17.5
Urban Interstates	1,644	3.8
Rural Interstates	1,983	4.5

NOTE: Estimates of fatalities on highways posted at 55 mph provided by the Center for Statistics and Analysis, National Highway Traffic Safety Administration, U.S. Department of Transportation.

fatality rates experienced in 1974 on rural Interstates; and (c) the effect of speed on safety has diminished in proportion to the reduced fatality rate between 1974 and 1982 on rural Interstates (15 percent).

As a result of increased speeds on rural Interstates, about 500 more fatalities would occur. This would increase deaths on highways posted at 55 mph by 2.3 percent and the number of deaths on rural Interstates by 20 to 25 percent.

Additional fuel would be consumed as a result of higher speeds. Approximately 10 million additional barrels of petroleum per year would be required. This would reduce the current savings in fuel consumption by 17 percent and would increase national consumption of petroleum by 0.18 percent (Table 52).

Although safety on rural Interstates would be impaired, if average speeds on the rural Interstate returned to 65 mph, travel on these high-

TABLE 52 Effect on Petroleum Consumption of 55 mph Speed Limit and Increased Speeds on Rural Interstates (*1*)

	Barrels Per Day (000s)	Percent of Total
National—all sectors	15,000	100
Transportation	9,200	57
Highway consumption	7,220	45
Consumption reduced by 55 mph speed limit	167	1.1
Impact of increased speeds on rural Interstates	27	0.18

ways would require 445 million fewer hours annually. This means that about 100 years of travel time would be saved for the loss of each additional life. On the remainder of the highway network the speed limit currently costs motorists about 600 million hours of additional travel time, or about 25 years of driving for every life saved. Considering the nationwide impact of the 55 mph speed limit, this means that the cost of the 55 mph speed limit, measured in additional travel time, is about four times as great on rural Interstate routes, per life saved, as it is on other highways posted at 55 mph. Thus, relaxing the 55 mph speed limit on these routes would have a proportionally greater effect on travel time than it would on safety.

These estimates, by necessity, are somewhat speculative. Some members of the committee believed that estimates of the impact of higher speeds on safety could not be accurately predicted. It has been 10 years since the speed limit had its pronounced impact on Interstate fatality rates, and many changes have taken place in the highway environment since that time. The limitation of aggregate statistics and the complexity of highway accidents precludes rigorous measurement of the benefits of the 55 mph speed limit on rural Interstate fatality rates. Despite this limitation, rough approximations of safety effects of the sort provided here provide the clearest means for communicating the scale of the probable safety effects of higher speeds. These approximations, however, are based on the assumption that any increases in the speed limit that are permitted would not affect other facilities. The validity and implications of this assumption are debatable, as discussed next.

Could Higher Speeds, If Allowed on Some Roads, Be Contained to Eligible Routes?

If speed limits higher than 55 mph were permitted on some roads, this could alter driver behavior on other roads as well. When the 55 mph speed limit first became effective in 1974, safety improved on almost all highways, even highway systems posted at speeds below 55 mph. Measures of speed trends in individual states indicate declines in average speeds and speed variance on highways posted at or below 55 mph before 1974. For example, in North Carolina, average speeds on highways without changes in posted speed limits declined 1 mph, and more important, the variability in speeds (as measured by the standard deviation) fell by one-half (Table 53). In short, the fastest slowed down the most.

The 55 mph speed limit may have contributed to these reduced speeds by changing drivers' perception of the appropriate safe speed on different highways. Perhaps as motorists came to associate 55 mph as a safe

TABLE 53 Comparative Speed Trends on Highways in North Carolina for Highways With and Without Changes in Posted Speed Limits (*6*)

		Pre-55 mph Speed Limit (April 1973)	Post-55 mph Speed Limit (April 1983)
Highways with reduced speed limit	Mean	54.0	51.9
	Standard deviation	3.2	2.58
Highways with no change in posted speed limit	Mean	50.5	49.3
	Standard deviation	4.89	1.93

speed on Interstates and highways of similar design quality, they unconsciously slowed down on other, less well-designed roads. Although this effect cannot be firmly proven, the 1974 experience suggests some validity to this hypothesis.

Similarly, if by increasing the speed limit on a few roads the federal government creates a public sense that the 55 mph speed limit on other roads is unimportant, slated for replacement, or unjustified, then the consequences of selectively increasing speed limits on a restricted group of roads could be greater than those described in the preceding paragraph. That is, authorization for speed increases on a small part of the highway system could produce a system-wide change in driving behavior. The strength of the spillover effect cannot be firmly established and might hinge on enforcement strategies adopted, but it does provide one more reason for caution in contemplating upward adjustments of the speed limit.

Is Control of Speed Limits a State or a Federal Responsibility?

Originally, Congress set the speed limit at 55 mph to save energy. The choice of 55 mph, rather than some other speed, emerged as a compromise. The President had proposed a speed limit of 50 mph to maximize fuel savings, but the trucking industry argued that private shipping would pay too high a price. Subsequently, the administration proposed 55 mph for trucks and 50 mph for automobiles, but Congress opted for the single speed limit of 55 mph.

In retrospect, the 55 mph national maximum speed limit, which began as a temporary measure, resulted in permanent changes in public policy that might not have been achieved otherwise. By requiring the states to conform to a national standard for maximum speeds, the federal government entered an area of policymaking previously occupied by state and local governments. State governments had generally accepted this shift amidst the national crisis precipitated by the OPEC embargo; however, when the law was made permanent as a safety policy, new concerns were raised.

Some practical considerations favor returning regulation of the speed limit to state and local governments. States and counties are more aware of local highway conditions than is the federal government. In the view of most committee members, however, this issue is overridden by the dramatic safety improvements that have resulted from the 55 mph speed limit. Immense practical difficulties had to be overcome to achieve the widespread behavioral change associated with the 55 mph speed limit, and extraordinary historical circumstances permitted this accomplishment. The majority of the committee, recognizing the unique context in which the current federal policy has evolved, believes it is appropriate for the federal government to continue to be involved in the determination of national maximum speeds.

CONCLUSION

The committee generally endorses the 55 mph speed limit as a national safety policy. Many thousands of lives have been saved during the last decade by slower and more uniform speeds on the highway. The committee strongly supports efforts to improve highway safety in order to reduce the tragic and unnecessary loss of life on the nation's highways. Some members believe that the 55 mph speed limit should be retained without exception; others believe that some moderate increases should be permitted on selected roads; and some believe that such increases should be permitted only if other safety policies are enacted simultaneously to compensate for the effect of higher speeds. The committee remained divided on whether such modification was necessary or desirable. Their division stems from divergent opinions on the following questions discussed earlier:

- Are the benefits of the 55 mph speed limit being eroded by higher speeds and other influences on safety?
- Will growing noncompliance lead to the nullification of this law?
- Does increased noncompliance with the 55 mph speed limit breed disrespect for law generally?
- What would consequences be of higher speeds on rural Interstates?

- Could higher speed limits, if allowed on some roads, be contained to eligible routes?
- Is control of speed limits a state or a federal responsibility?

Many of the questions and issues involved extend beyond those answerable by scientific and technical analysis, and the committee believes that it can most productively contribute to an informed resolution of this issue by presenting the various facts and arguments that relate to these questions, as has been done in this chapter.

In addition, the committee recommends that the current standard for measuring state compliance be replaced. Adjustments to state speed data should be based on accepted principles of speed-limit enforcement, statistically valid measurement error should be accounted for, and all states should be treated as fairly as possible. By adopting a point system in which high-speed violations are weighted more heavily than incremental violations, the measurement system could be brought more in line with accepted social behavior and would improve safety by encouraging state patrols to focus enforcement on speeding at high-risk locations (Chapter 9).

Recommending some future course of policy relative to the 55 mph speed limit necessarily involves judgments about trade-offs between the various consequences, as well as judgments about many other issues that surround this choice. These trade-offs ultimately require judgments about uncertain consequences, the value of these consequences, and the extent to which various governmental actions can or should preserve them. These choices, based on values that individuals in a pluralistic society do not find of equal importance, can only be made by the elected representatives of the American public.

The committee strongly believes that in considering alternatives to the current policy, Congress should be aware of the substantial consequences that have resulted from the 55 mph speed limit, the unique circumstances under which this change in policy was achieved, and the costs and risks associated with any major modifications.

REFERENCES

1. *Monthly Energy Review*. Energy Information Administration, U.S. Department of Energy, Dec. 1983.
2. *The Economic Cost to Society of Motor Vehicle Accidents*. NHTSA, U.S. Department of Transportation, 1983.
3. Gallup Poll 1982. *The Gallup Report: Political, Social and Economic Trends*. Report 205. Princeton, N.J.
4. R. Berger and G.S. Persinger. *Survey of Public Perceptions on Highway*

Safety, 1980. HS-805-702. NHTSA, U.S. Department of Transportation, 1980.
5. J. Mounce. Driver Compliance with Stop Sign Control at Low Volume Intersections. In *Transportation Research Record 808*. TRB, National Research Council, Washington, D.C. 1981.
6. F. Council, L. Pitts, M. Sadorf, and O. Dart. *An Examination of the Effects of the 55 mph Speed Limit on North Carolina Accidents*. Highway Safety Research Center, University of North Carolina, Chapel Hill, Aug. 1975.

Appendix A
Effect of the Fuel Shortage on Recreational Travel in 1974

With the onset of the fuel shortage in late 1973 many people postponed vacations and reduced recreational trips. Vacation travel, especially long-distance driving on unfamiliar roads, may be more hazardous than routine trips for work or shopping. Unfortunately, the purposes of trips made by victims of accidents are not systematically recorded. National traffic data do not distinguish between recreational trips and work trips. The available evidence on recreational travel, however, supports the argument that, at least in some states, discretionary trips declined more than the estimated decline in total nationwide highway travel.

Several states have automatic traffic counters on highways leading to recreational or tourism destinations. Although not all traffic on these roads is recreational, the results can be used to approximate recreational travel trends. Statistics on visits to primary tourism sites can also be used to estimate recreational highway travel. Other states record out-of-state licenses to forecast tourist traffic. Several states were able to supply the data that they use in making recreational travel estimates. Information provided by Florida, New Mexico, Washington, and Wisconsin indicate that some states experienced declines in recreational travel greater than declines in total travel.

FLORIDA

The energy crisis of 1974 represented a clear break in the long-term trend in recreational travel growth in Florida. Recreational travel declined by 4.8 percent in 1974, only slightly more than the statewide travel decline of 4.65 percent. Motor vehicle fatalities declined by 17.3 percent, or by about the national average. But the annual data mask the impact of the energy crisis on travel patterns. The entire decline in travel in 1974 occurred during the first 6 months when fuel supplies were most scarce. Travel in the second half of the year was actually higher than during the corresponding period in 1973. Highway deaths declined more sharply in the first two quarters of 1974 than in the last half of the year. However, even after travel recovered, motor vehicle fatalities remained 10 percent below comparable periods in 1973 (Tables A-1 and A-2).

NEW MEXICO

Overall travel was virtually unchanged in 1974, down less than 0.3 percent. However, the number of out-of-state vehicles in New Mexico declined about 10 percent, and the number of visitors to national parks declined 16 percent. Fatalities on New Mexico's roads declined 15 percent in 1974. The decline in tourist travel is more in accord with the decline in highway deaths than the change in total statewide travel.

WASHINGTON

Information is collected on visits to state parks and house-trailer traffic. The state also monitors travel volumes along selected recreational routes. Monthly visitation data with quarterly and annual summaries, along with a monthly comparison of travel and fatalities, are given in Table A-3.

As in other states the decline in travel in Washington was concentrated in the first 5 months of 1974, as was the decline in highway deaths. The decline in vacation travel in early 1974 is even more pronounced. House-trailer traffic almost disappeared in the first quarter of 1974—declining by 57.3 percent. For the year, however, total counts of visitors traveling by house trailer increased 21.5 percent. Vacationers may have simply postponed their trips for later in the year (Table A-4).

Similar findings hold for data from counters on roads leading to recreational areas. For the first quarter, traffic for the average day was down 20.5 percent, but it was only down 11.3 percent for the second quarter. The third quarter was unchanged between 1973 and 1974 and a 10.1 percent increase occurred in the fourth quarter. Traffic on these roads

TABLE A–1 Number of Tourists Traveling by Car in Florida, Quarterly Data

Quarter 1973	No. of Tourists (000s)	Percent Change From Previous Year	Quarter 1974	No. of Tourists (000s)	Percent Change From Previous Year	Quarter 1975	No. of Tourists (000s)	Percent Change From Previous Year
1st	4,308	+10.3	1st	3,497	−18.8	1st	4,993	+42.8
2nd	4,708	+10.9	2nd	4,213	−10.5	2nd	4,325	+2.7
3rd	4,671	+0.01	3rd	5,000	+7.0	3rd	5,250	+5.0
4th	4,447	+16.4	4th	4,550	+2.3	4th	4,778	+4.9

TABLE A–2 Quarterly Decline in Fatalities and Tourism, 1974

Quarter	Percent Change in Tourism	Percent Change in Fatalities
1st	−18.8	−20.5
2nd	−10.5	−15.7
3rd	+7.0	−11.7
4th	+2.3	−10.5

TABLE A–3 Washington State Tourism Data, Monthly Change in Fatalities and Tourism Travel, 1973–1974

Total Recreational Travel (000s)			Percent Change in Tourism	Percent Change in Fatalities
Month	1973	1974		
January	616	510	−17.2	−6.0
February	899	588	−34.6	−7.0
March	1,207	930	−23.0	−13.8
April	1,705	1,665	−2.4	−52.1
May	3,281	2,961	−9.8	−30.3
June	3,457	4,507	+30.4	+4.1
July	5,452	5,505	+1.0	+11.8
August	4,852	6,059	+24.9	−10.0
September	2,266	2,610	+12.9	+4.8
October	954	1,277	+32.9	+34.4
November	527	667	+26.6	+22.4
December	429	547	+27.5	+17.0
Total	23,645	27,526		

TABLE A–4 House Trailer Visits to Washington State Parks

Month	1973	1974	Percent Change
January	2,573	1,189	−53.8
February	14,577	5,218	−64.2
March	26,948	5,578	−79.3
April	60,608	43,611	−28.0
May	80,511	92,338	+12.8
June	127,675	184,475	+44.5
July	234,670	267,822	+14.2
August	231,925	322,787	+39.2
September	72,039	126,208	+75.2
October	21,478	48,384	+125.3
November	6,115	11,747	+92.1
December	2,611	4,200	+61.1
Total	879,730	1,120,422	+21.5

was down 27.9 percent in the first quarter for Saturday and Sunday. Overall the Washington state data indicate that recreational travel declined more steeply than total travel. Total vehicle miles of travel (VMT) declined only 3.7 percent statewide.

WISCONSIN

Data on visits to state parks are recorded but no direct estimates of recreational automobile traffic are made. Total state traffic declined 2.6 percent in 1973, and the number of visitors to state parks declined by 3.7 percent. However, neither approaches the 20 percent decline in motor vehicle fatalities recorded in Wisconsin in 1974. The Wisconsin data do not support the thesis that changes in the composition of highway travel explain much of the improved highway safety record in 1974.

Appendix B
Effectiveness of Safety Belts in Reducing Highway Fatalities

Safety belts have been in existence nearly as long as automobiles. The first patent for leather straps to restrain the occupants of a motorized vehicle dates to 1885. Until recently, however, research into the effectiveness of safety belts in reducing injuries has been limited to their use in air transportation. As early as 1915 aircraft designers encouraged pilots to wear restraints to minimize the risk of being killed when thrown from a crash. Later research supported by the U.S. Air Force demonstrated the ability of the human body to withstand remarkably high forces of deceleration when adequately restrained in a belt and harness (*1*).

In the first large-scale analysis of automobile crashes it was estimated that some 20 percent of automobile crash injuries was caused by ejection of the occupant, and the use of safety belts was encouraged to minimize this risk (*2*). Since that report dozens of studies have documented the effectiveness of safety belts in reducing the severity of injuries. Although the role of safety belts in minimizing ejection continues to be important, especially in rollover crashes, more recent research has emphasized the importance of minimizing the force of the second impact—the buffeting of the occupants against the inside of the car—that occurs in almost all crashes.

Most studies of safety belt effectiveness have relied on statistical analyses that compare the probability of severe injury for belted as opposed to unbelted occupants. A few studies, however, have applied a clinical, or case study approach in which medical researchers reconstruct crashes to estimate whether a belted occupant would have had a less severe injury. The findings from these two approaches result in a fairly broad range of effectiveness in reducing serious injury or fatality. Statistical analyses estimate the effectiveness of safety belts to range between 50 and 90 percent. Clinical analyses estimate belts to be about 30 percent effective.

STATISTICAL ANALYSES

Statistical approaches that estimate the effectiveness of safety belts in reducing the risk of fatality, as opposed to injury, generally have been hampered by small samples. In order to have a sufficient number of observations, most studies have combined the number of severe injuries with fatalities and then compared the frequency of these injuries for belted and unbelted occupants. In a comprehensive review of the literature through 1972, Griffin (*3*) found that safety belts reduced the most severe injuries and fatalities by 35 to 65 percent.

The most recent studies of the effectiveness of safety belts have been aided by expanded data bases, which have allowed the analysis to control for the confounding influences of vehicle size and weight, driver age, type of accident, and accident severity. For example, Reinfurt et al. (*4*) found that three-point belts (combined lap-shoulder variety) reduced injuries for all types of accidents and in all sizes of vehicles (Table A-5). Safety belts were found to be almost as effective in the most severe, highest speed crashes, as in minor ones. This study, despite a sample of more than 15,000 occupants involved in crashes, did not have a sufficient number of fatal accident victims to estimate the effectiveness of safety belts on fatalities.

In a review of the literature on safety belts, Campbell and Reinfurt (*5*) noted the problems involved in determining the effectiveness of safety belts in reducing fatalities, but they also pointed out a fairly high degree of consistency in the findings of studies of three-point belts. Part of the variance in the estimates for the combination of serious and fatal injuries results from different definitions of serious injury. Even so, most studies reveal a rising level of effectiveness as the measure of injury severity increases (Table A-6). Campbell and Reinfurt (*5*) suggest that because of this consistency across studies, the estimates of reductions in fatalities of 75 percent or greater may not be unrealistic.

TABLE A–5 Effectiveness of Lap-Shoulder Safety Belts in Reducing Moderate and Severe Injuries (*4*)

	Injury	
	Moderate or Worse (%)	Severe or Worse (%)
Vehicle Damage		
Minor	53–56	43–46
Moderate	58–60	65–66
Moderate/severe	46–55	36–62
Severe	44–51	39–49
Age of Occupant		
10–25	48	50
26–55	53	41
56+	59	53
Crash Type		
Front	39	–[a]
Side	58	–[a]
Rear	61	–[a]
Rollover	65	–[a]
Vehicle Type		
Subcompact	52	–[a]
Compact	52	–[a]
Intermediate	45	–[a]
Full-sized	60	–[a]

[a]Insufficient sample size for estimation.

CLINICAL ANALYSES

A less frequently applied method of analyzing the effectiveness of safety belts has relied on expert judgment. Wilson and Savage (*25*) pooled expert opinion on the probable outcomes of 706 fatal accidents had safety belts been worn. They estimated that the combination of shoulder and lap belts would have reduced the number of fatalities by 30 percent. In a more recent analysis of fatal accidents in mostly rural, high-speed crashes, Huelke et al. (*13*) estimated that about 30 percent of lives could have been saved if safety belts had been worn. In another study of a wider variety of crashes, Huelke et al. (*12*) found safety belts much more effective in reducing fatalities (Table A-6).

A review of the safety-belt effectiveness literature by O'Day and Flora (*26*) suggests that an overall effectiveness of about 30 percent "may be closest to the truth." They point out that drivers who wear safety belts

TABLE A–6 Safety Belt Effectiveness Studies (*5*)

Author	Year	Type of Belt	Reduction in All Types of Injuries (%)	Reduction in Serious and Fatal Injuries (%)	Reduction in Fatalities Only[a] (%)
Anderson (*6*)	1972	Lap	–5[b]	8	7.5
Andreassand (*7*)	1969	Three point	23	45	–[d]
Andreassand (*7*)	1972	Three point	26	30	72
Bohlin (*8*)	1973	Three point	36	44	83
Campbell (*9*)	1968	Lap	–[d]	34	–[d]
Commonwealth Bureau of Statistics (*10*)	1973	Three point	17	–[d]	54
Council and Hunter (*11*)	1974	Combined	39	–[d]	53
Huelke (*12*)	1977	Three point	–[d]	47–57	77–91
Huelke[c] (*13*)	1979	Three point	64	–[d]	32
Highway Safety Foundation (*14*)	1970	Lap	–[d]	56	81
Kilhberg (*15*)	1969	Lap	30	54	26
Levine and Campbell (*16*)	1971	Lap	36	48	–[d]
New York State DVM (*17*)	1972		–[d]	–[d]	86
Preston and Shortridge (*18*)	1973		12	42	–[d]
Reinfurt, et al. (*4*)	1976	Lap	31–39	46–53	
		Three point		57–62	
Richardson (*19*)	1972	Lap	51–68	36	59
Rinninger and Boak (*20*)	1976	Three point	–[d]	69	77
Road Trafffic Board (*21*)	1972	Three point	40		68
Scott (*22*)	1975		46	–[d]	–[d]
Tourin and Garret (*23*)	1960	Lap	–5[b]	27	38
Washington State Patrol (*24*)	1970	Lap	20	67	80
Wilson and Savage[c] (*25*)	1973	Three point	–[d]	–[d]	31

[a]Given the small number of fatalities for belted occupants in most studies, these estimates should be interpreted with caution.
[b]The negative sign indicates an increase in injuries. These increases occurred for relatively minor injuries, perhaps due to improper safety belt adjustment.
[c]Clinical, rather than statistical, methodology.
[d]Not estimated is this category.

are only about one-half as likely to be involved in the most severe crashes. They compared the probability of accident involvement for belted and unbelted drivers (*26*) for accidents at all levels of severity (Figure A-1). The traffic accident data (TAD) measure used in Figure A-1 is an estimate made by the investigating officer of the amount of vehicle crush. Drivers who wear safety belts are only about one-half as likely to be involved in the most severe crashes. This suggests that people who wear safety belts are safer, more cautious drivers. When involved in accidents, they simply do not hit objects as hard. O'Day argues that if the lower likelihood of severe accident involvement by drivers wearing safety belts is corrected, safety belts reduce fatalities by about 30 percent.

Statistical and clinical analyses can assess the potential benefits of safety belts, but the extent to which that potential can be realized depends on the extent of safety belt use, particularly in situations where serious

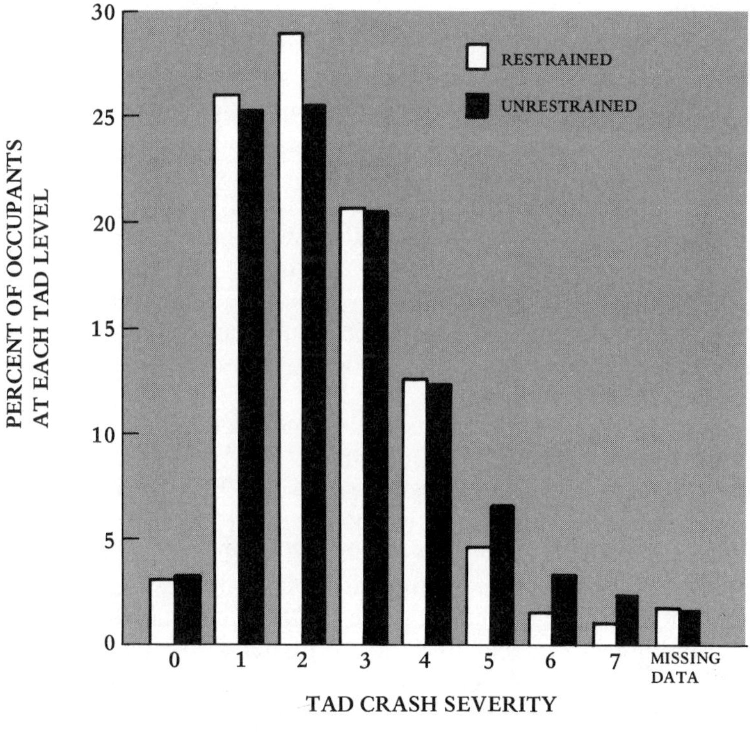

NOTE: TAD refers to traffic accident data. A rating of TAD 7 is the most severe.

FIGURE A–1 Crash severity for belted and unbelted drivers (*26*).

accidents are most likely to occur. Motorists may not realize these potential safety benefits, as is evident by the low percentage of U.S. motorists that wear safety belts. Alternatively, given the low probability of being involved in an accident, the non-use of safety belts may indicate a willingness to risk a more severe injury to avoid the inconvenience of wearing a safety belt. Passage of mandatory safety belt laws may not yield the full potential benefit, because noncompliance in general, and among high-risk groups in particular, can greatly diminish the benefits actually obtained. Thus a realistic assessment of the maximum benefits of safety belts must come from the reductions in fatalities and serious accidents that have actually been achieved after governments have mandated the use of safety belts.

INTERNATIONAL EXPERIENCE WTH MANDATORY-USE LAWS

Since the early 1970s more than 30 nations have passed mandatory-use laws (Table A-7). Victoria, Australia, began requiring the use of safety belts in 1970, and by 1972 the remaining states in Australia and New Zealand had followed suit. Most European nations began requiring use of safety belts during the middle part of the decade, as did Puerto Rico. Ontario passed the first mandatory-use law in Canada in 1976 and was rapidly followed by Quebec, Saskatchewan, and British Columbia, and recently by Newfoundland and New Brunswick. These laws apply to all vehicles equipped with safety belts and to all types of driving. Laws passed in Japan, Israel, and Spain limit the requirement for safety belt use to intercity or expressway travel. In January 1983 Great Britain began requiring mandatory use of safety belts after debating the issue throughout the 1970s.

In most nations motorists responded positively to the requirement that they wear safety belts. For example, safety belt use in Victoria increased nearly 60 percentage points, from only 20 to almost 80 percent. Because few nations vigorously enforce their mandatory-use laws, compliance appears to depend on a relatively high level of public acceptance. In the few cases where surveys of public opinion were taken before passage of the mandatory-use law, the majority of motorists favored the idea even though few actually wore their belts at the time.

Little information is available to estimate the level of compliance over the long term, and the evidence is mixed. Estimates of short-term use are available for about one-half of the countries that have mandatory-use laws, as are longer-term estimates for about one-third. In some nations, such as Australia, New Zealand, Norway, Sweden, and West

TABLE A–7 International Mandatory-Use Laws and Estimates of Use (*27–30*)

Jurisdiction	Date Of Law	Type[a] Enforce-ment	Pre-Law Use: Urban (%)	Pre-Law Use: Rural (%)	Pre-Law Use: Combined Urban-Rural (%)	Post Law Use: Urban (%)	Post Law Use: Rural (%)	Post Law Use: Combined Urban-Rural (%)	Most Recent Estimate: Date	Most Recent Estimate: Urban (%)	Most Recent Estimate: Rural (%)	Most Recent Estimate: Combined Urban-Rural (%)
Victoria, Australia	1970	2	—	—	18	75	64	—	1981	90	80	—
Austria	1976	0	5–10	25	—	15	40	—	—	—	—	—
Belgium	1976	2	—	—	17	—	—	85	—	—	—	—
British Columbia	1977	3	—	—	35	—	—	45	1982	—	—	50
Ontario	1976	3	—	—	25	—	—	25	1982	—	—	50
Quebec	1976	3	—	—	18	—	—	75	1982	—	—	70
Saskatchewan	1977	3	—	—	15	—	—	55	1982	—	—	45
Denmark	1976	2	14	34	—	79	84	—	—	—	—	—
Finland	1975	—	8	31	—	38	66	—	—	—	—	—
France	1979	2	—	—	20–25	30–35	75–85	—	—	—	—	—
Great Britain	1983	0	—	—	38–54	—	—	95	1983	—	—	95
Israel[b]	1975	—		24			75	—	1977		70	
Luxemburg	1975	—	8	31	—	53	66	—	—	—	—	—
Netherlands	1975	0	11	24	—	58	75	—	—	—	—	—
New Zealand	1972	3	33	52	—	87	91	—	1979	—	—	85
Norway	1975	0	15	38	—	30	65	—	1978	30	65	—
Puerto Rico	1974	0	—	—	3	—	—	25	1983	—	—	6
Sweden	1975	—	20	35	—	80	85	—	1978	80	85	—
Switzerland	1980	—	19	35	—	75	81	—	1978	30	50–60	—
West Germany	1976	0	25	70	—	45	80	—	1980	45	80	—

NOTE: For the following jurisdictions, only information on the date of passage was available: the other provinces of Australia, 1970–1972; Brazil, 1977; Bulgaria, 1976; other Canadian provinces: Newfoundland, 1982; New Brunswick, 1983, and Manitoba, 1984; Czechoslovakia, 1975; Greece, 1979; Hungary, 1977; Iceland, n.d.; Ireland, 1979; Ivory Coast, 1970; Malawi, 1970; Portugal, 1978; South Africa, 1977; Spain, 1974; Turkey, n.d.; Yugoslavia, 1977; USSR, 1975.

[a]Key: 0 = no enforcement; 2 = enforced in conjunction with other laws; 3 = intermittent enforcement; dash = no data available.
[b]Law applies in rural areas only.

Germany, the initial increases in compliance appear to have been sustained. Whereas New Zealand relies on intermittent enforcement, West German drivers appear to wear safety belts voluntarily. There is no fine for non-use and the law is not enforced (non-use may affect compensation in the event of litigation). In contrast, in Puerto Rico use of safety belts never exceeded 25 percent and is even lower today. In Canada use of safety belts increased to nearly 75 percent in the immediate aftermath of the law, but use has recently been estimated at about 50 percent (*31*).

Estimates of Effectiveness

Few nations have analyzed the effectiveness of their mandatory-use laws in reducing injuries and fatalities. The limited research that has been done indicates that the effects were not as large as expected. For example, despite the increase in compliance of nearly 67 percentage points in New Zealand, fatalities declined only about 12 percent in the first year the law was passed (*32*). Similar results were reported in Switzerland (*33*). A review of Denmark's experience revealed that the number of casualties declined in the first year the mandatory-use law was passed, but no effect could be found in the second year (*34*). The exception to these reports has been the recent experience in Great Britian. After passage of a mandatory-use law, the rate of safety belt use increased from less than 50 to more than 90 percent, whereas fatalities decreased 25 percent. This experience indicates the effectiveness of safety belts in reducing fatalities to be 50 percent (*35*).

Closer to the United States geographically, a recent comparison of the Canadian provinces with and without mandatory-use laws indicates quite similar improvements in occupant fatality rate trends, despite the low percentage of safety belt use in the provinces without mandatory-use laws (*31*). Jonah and Lawson (*31*) estimated the benefits of the mandatory-use laws by calculating the greater reduction in occupant fatalities in provinces with mandatory-use laws than those without such laws. Using this technique they estimated that Ontario's mandatory-use law contributed to a 13.7 percent greater reduction in fatalities than in the provinces without mandatory-use laws (while noting the possible interaction of safety belts with the reduced speed limits introduced at the same time). In comparison, Quebec experienced only a 3.5 percent greater reduction in fatalities. On the basis of the authors' assumption that safety belts are 60 percent effective in reducing fatalities, reductions of 18 percent in Ontario and 14 percent in Quebec were expected.

The reason for these smaller than expected effects may result from the unwillingness of the least safe drivers to wear safety belts (*32,36,37*).

Rates for use of safety belts are typically much lower for teenagers, drivers at night, and drivers who have been drinking, all of whom have a higher probability of involvement in an accident. This implies that the effectiveness of safety belts does not increase linearly with compliance. The effectiveness of safety belts in Great Britian—where almost all drivers are reported to wear their belts—supports this point. With virtually universal use of safety belts, their effectiveness in practice appears to be in line with the predictions from statistical analyses. As O'Day and Flora noted (*26*), motorists who wear safety belts are less likely to be involved in the most severe accidents yet these drivers will be the first to "buckle up." If this is true, compliance with the law would have to be extremely high to obtain the full potential benefits of safety belts.

ESTIMATE OF THE EFFECTIVENESS OF A MANDATORY SAFETY-BELT USE LAW IN THE UNITED STATES

Estimating the number of lives that might be saved in the United States if a mandatory-use law is passed depends, in large measure, on the expected level of compliance. The available evidence suggests that compliance would not be high. Only about 11 to 14 percent of Americans wear safety belts today, an actual decrease since 1970. Further, a 1982 poll indicates that only 19 percent of the public support a mandatory-use law (enforced with a $25 fine) compared to 23 percent in 1973 (*38*). As noted previously, in the nations that have passed mandatory-use laws, most of the public supported the idea before passage.

In contrast to this bleak picture of public acceptance, more than 40 states have recently enacted mandatory child-restraint laws. Many safety belt advocates interpret this as a sign of growing acceptance of mandatory restraints for all drivers. Further, in almost all nations that passed mandatory-use laws, large-scale public education programs had been underway, sometimes for many years. Although these programs may have had a minimal impact on the use of safety belts, they may have shifted public opinion about the potential benefits of the law. A similar shift in opinion might occur in the United States with a large-scale public education effort.

Estimated Benefits

The benefits to the United States of a mandatory-use law depends on the base of comparison. The experience in Great Britain indicates that if virtually all motor vehicle occupants wear safety belts, the risk of fatality is reduced by about 50 percent. Translated to the United States, this could mean a dramatic reduction in the number of vehicle-occupant fatalities. More than 12,000 lives could be saved. The fundamental question is whether Americans will wear safety belts. Although the answer will not be known until a mandatory-use law goes into effect, the available indicators are not promising. Public support for a mandatory-use law is low. Few motorists elect to wear their safety belts voluntarily. And experience with speeding and drunken driving demonstrates the difficulty of changing public behavior through law. Although many more motorists would wear safety belts in the wake of the notoriety of a mandatory-use law, as they have in other nations, universal use of safety belts appears unlikely.

For this reason the U.S. experience is more likely to parallel that of the Canadian provinces. If safety-belt use rates increased about 50 percentage points, and assuming an effectiveness rate equal to that experienced in Ontario, between 4,000 and 6,500 lives might be saved. Injury severity would be reduced in thousands of accidents as well. On the other hand, if U.S. experience paralleled that of Quebec, the reductions in fatalities might be smaller. If only the relatively safe drivers comply with the law the reductions in fatalities among this group would be on the order of 750 to 1,000.

SUMMARY

For individual motorists, the wearing of safety belts significantly reduces the risk of serious injury or fatality. Even though the risk of involvement in a crash is small, when these small risks are spread over the entire population, crashes occur in large numbers and thousands of motorists die each year. By wearing safety belts these individuals could have reduced their risk of being killed. Because most automobiles in the United States are already equipped with safety belts, the costs of a mandatory safety-belt use law would be small.

As indicated in this review of the experiences of other nations, the benefits of a mandatory-use law will vary widely depending on how the public responds. Although the most optimistic projections of public compliance would result in substantial savings in lives, the benefits could be much smaller. Compliance will be influenced by factors difficult to predict. Historic patterns of use indicate that few Americans wear safety belts voluntarily. The same can be said of other motorists in other nations. However, despite low-use rates, use increased dramatically in other nations with passage of a mandatory-use law. Public acceptance of the law depends on the treatment of the topic in the media and the awareness gained through public information in publicity campaigns. Campaigns underway at the national, state, and local levels could help convince individuals of the merits of safety belts and increase support for a law mandating their use. On the other hand, the role of government in shaping personal behavior is much more constrained in the United States than in other nations. Whether the public would accept an expanded governmental role, regardless of the safety benefits, remains to be seen. Recent passage of a mandatory-use law in New York state (to become effective January 1, 1985) will provide an important test of public acceptance and safety belt effectiveness.

One of the policy options discussed by the committee would permit states to increase speed limits on selected highways if other compensating safety policies were implemented. A mandatory safety-belt use law ranks high among available policies that states would easily implement to reduce highway fatalities. Since the committee began its discussions, mandatory safety-belt use laws have been linked to another safety policy debate. In her final ruling on Federal Motor Vehicle Safety Standard 208 (the federal regulation requiring airbags or automatic safety belts), Secretary of Transportation Elizabeth Hanford Dole ruled that if states representing two-thirds of the population passed laws requiring the use of safety belts by 1990, the rule requiring motor vehicle manufacturers to install passive restraints would be waived. This policy decision effectively preempts the option of mandatory safety-belt use policies as a quid pro quo for higher speed limits.

REFERENCES

1. L. Ronan. *Seat Belts*: *1949-1956*. NHTSA, U.S. Department of Transportation, 1979.
2. J. Moore and B. Tourin. *Study of Automobile Doors Opening Under Crash Conditions*. Automotive Crash Research Project, Moore Collections, Cornell University, Ithaca, N.Y., 1954.
3. L.I. Griffin, III. *Analysis of the Benefits Derived from Certain Presently Existing Motor Vehicle Safety Devices*: *A Review of the Literature*. Highway Safety Research Center, University of North Carolina at Chapel Hill, 1973.
4. D.W. Reinfurt, C.Z. Silva, and D.F. Seila. *A Statistical Analysis of Seat Belt Effectiveness in 1973-1975 Model Cars Involved in Towaway Crashes*. University of North Carolina at Chapel Hill, 1976.
5. B.J. Campbell and D.W. Reinfurt. *The Degree of Benefit of Belts in Reducing Injury—An Attempt to Explain Study Descrepancies*. SAE Technical Paper 790684. Society of Automotive Engineers, Warrendale, Pa., 1979.
6. T.E. Anderson. *Analysis of Vehicle Injury Sources*, Final Report. Cornell Aeronautical Laboratory, Buffalo, N.Y., 1972.

7. D.C. Andreassand. The Effects of Compulsory Seat Belt Wearing Legislation in Victoria. Presented at the National Road Safety Symposium, Canberra, Australia, 1972.
8. N.I. Bohlin, H. Norm, and A.A. Anderson. *A Statistical Traffic Accident Analysis with Reference to Occupant Restraint Value and Crashworthiness of the Volvo Experimental Safety Car*. Gothenburg, Sweden, 1973.
9. B.J. Campbell. *Seat Belts and Injury Reduction in 1967 North Carolina Automobile Accidents*. Highway Safety Research Institute, University of North Carolina, Chapel Hill, 1968.
10. *Report on a Statistical Investigation into the Effectiveness of Seat Belts in Motor Vehicle Accidents*. Commonwealth Bureau of Statistics, Victoria Office, Sydney, Australia, 1973.
11. F.M. Council and W.W. Hunter. *Seat Belt Usage and Benefits in North Carolina Accidents*. Highway Safety Research Center, University of North Carolina, Chapel Hill, 1974.
12. D.F. Huelke, T.E. Lawson, R. Scott, and J.C. Marsh. *The Effectiveness of Belt Systems in Frontal and Rollover Crashes*. SAE International Automotive Engineering Congress and Exposition, Detroit, Mich., 1977.
13. D.F. Huelke, H.W. Sherman, M.J. Murphy. *Effectiveness of Current and Future Restraint Systems in Fatal and Serious Injury Automobile Crashes*. SAE International Automotive Engineering Congress and Exposition, Detroit, Mich., 1979.
14. Highway Safety Foundation. *A Study of Seat Restraint Use and Effectiveness in Traffic Conditions*. Mansfield, Ohio, 1970.
15. J.K. Kihlberg. *Efficacy of Seat Belts in Injury and Non-Injury Crashes in Rural Utah*. Final Report. Cornell Aeronautical Laboratory, Buffalo, 1969.
16. D.N. Levine and B.J. Campbell. *Effectiveness of Lap Seat Belts and the Energy Absorbing Steering System in the Reduction of Injuries*. Highway Safety Research Center, University of North Carolina, Chapel Hill, 1976.
17. New York State Department of Motor Vehicles. Final Report. VSDSS Research Studies. Albany, 1973.
18. F.L. Preston and R.M. Shortridge. *A Study of Restraint System Use and Effectiveness*. University of Michigan Highway Safety Research Institute, Ann Arbor, 1973.
19. H.A. Richardson. *Statistical Analysis of Safety Belt Usage in the State of Oregon*. Preliminary Report. NHTSA, U.S. Department of Transportation, 1972.
20. A.R. Rininger and R.W. Boak. "Lap-Shoulder Belt Effectiveness." *Proc., 20th Annual American Association for Automotive Medicine Conference*, Atlanta, Ga. 1976.
21. Road Traffic Board of South Australia. *A Review of Seat Belt Usage in South Australia, 1964–1971*. Canberra, Australia, 1972.
22. R.E. Scott. Progress Report, MVMA Contract Study, March 1974–April 1975. University of Michigan Highway Safety Research Institute, Ann Arbor, 1975.
23. B. Tourin and J.W. Garrett. *Safety Belt Effectiveness in Rural California Automobile Accidents: A Comparison of Injuries to Users and Non-Users of Safety Belts*. Automotive Crash Injury Research of Cornell University, Buffalo, N.Y., 1960.
24. Washington State Patrol. *Seat Belt Study, 1970*. Olympia, 1971.

25. R.A. Wilson and C.M. Savage. *Restraint System Effectiveness: A Study of Fatal Accidents*. General Motors Automotive Safety Engineering Seminar, June 20-21, Warren, Mich., 1973.
26. J. O'Day and J. Flora. *Alternative Measures of Restraint System Effectiveness: Interaction with Crash Severity Factors*. SAE Technical Paper 820798. Society of Automotive Engineers, Warrendale, Pa., 1982.
27. G.M. Mackay. "Seat Belts in Europe—Their Use and Performance in Collisions." *Proc., International Symposium on Occupant Restraint*, June 1-3, American Association for Automotive Medicine, Morton Grove, Ill., 1981.
28. F.G. Fisher, Jr. *The Effectiveness of Safety Belt Usage Laws*. NHTSA, U.S. Department of Transportation, 1980.
29. S.J. Ashton, G.M. Mackay, and S. Camm. "Seat Belt Use in Britain under Voluntary and Mandatory Conditions." *Proc., 27th Annual American Association for Automotive Medicine, Conference*, Arlington Heights, Ill., 1983.
30. A. Grimm. Chart of Jurisdictions with Restraint Use Laws, unpublished, available from University of Michigan, Transportation Research Institute Library, Ann Arbor, 1984.
31. B. Jonah and J. Lawson. *The Effectiveness of the Canadian Mandatory Safety Belt Use Laws*. Road Safety Directorate, Transport Ottawa, Canada, 1983.
32. P.M. Hurst. "Compulsory Seatbelt Use: Further Inferences." *Accident Analysis and Prevention*, Vol. 11, Pergamon Press, Great Britain, 1979.
33. K. Hell. Injuries of Automobile Occupants in Traffic Accidents with and without the Seat Belt Requirement. Ver von prospehtiven 1, Jahresstudien uber Autonfelle innrouts im Katon Basel Stadt, May 1977 as cited in Fisher (1980).
34. E. Nordentoft, T. Kruse, H. Neelsen, and R.P. Weelk. "The Effect of Mandatory Seat Belt Legislation on Mortality and Morbidity in Denmark." *Proc., 22nd Annual American Association for Automotive Medicine Conference*, Melbourne, Australia, as cited in Fisher (1980).
35. G.M. Mackay. "Consequences of Mandatory Safety Belt Usage in Britain." Presented at 63rd Annual Meeting of the Transportation Research Board, Washington, D.C., 1984.
36. L. Evans and P. Wasielewski. "Risky Driving Related to Driver and Vehicle Characteristics." *Accident Analysis and Prevention*, Vol. 15, 1983.
37. L. Robertson."Automobile Seat Belt Use in Selected Countries, States, and Provinces With and Without Laws Requiring Use." *Accident Analysis and Prevention*, Vol. 10, 1978.
38. The Gallup Poll 1982. *The Gallup Report: Political, Social, and Economic Trends*. Report 205, Princeton, N.J., Oct., 1982

Appendix C
Speeding and Highway Fatalities

SUMMARY AND RESULTS

This appendix was written by Charles A. Lave, Professor of Economics, University of California, Irvine. The multiple regression analysis used in this appendix compares fatality rates at the state level to state-level measures of vehicle speeds, availability of medical treatment, and enforcement activity based on 1981 and 1982 data. Cross-sectional regression analysis is limited insofar as it is unable to control for all the potentially important variables affecting the dependent variable. Nonetheless, the relationship between fatality rates and speed variance and the surprising lack of relationship between fatality rates and average speeds described in this analysis, further suggest the need both to improve the quality of data and to extend the research in this area. The following conclusions emerged from this regression analysis.

First, analysis of the 1981–1982 data reveals no statistically significant relationship between average speed and the fatality rate. This is true for other speed measures as well: percentage of drivers exceeding 55 mph, percentage of drivers exceeding 65 mph, and 85th percentile speed. All considered, states with high average speeds do not have higher fatality rates than states with low average speeds.

Second, there is a statistically significant relationship between speed variance (the range of speeds on some given highway) and the fatality rate. When most cars are traveling at about the same speed, whether it is a high speed or a low speed, the fatality rate is low; presumably because the probability of collision is low. When there is a considerable range of speeds among cars on the highway, the fatality rate is high; presumably because this increases the probability of collision.

Third, when the effect of speed variance is held constant, there is no statistically significant relationship between the fatality rate and any other speed variable. This suggests that the variance of speed is more important to safety than average speed, but the policy implications of this relationship are unclear. First, it is not clear whether the variance of speed can be controlled independently of the average. Second, it is uncertain whether results based on cross-sectional differences will accurately predict the effects of larger differences outside the range of cross-sectional variation. Nevertheless, these results suggest that measures to control the variance of speed could be an effective component of traffic control. This might mean curbing those drivers whose speed is substantially different from that of the traffic flow (slower as well as faster). A point system, similar to the one described in Chapter 9, is a change in this direction because it encourages police to concentrate on the extreme speeders.

DATA AND METHODOLOGY

This study used multiple regression analysis to predict the variable "fatalities per vehicle mile traveled." Data points were state average figures; that is, there were 50 potential data points for each regression. Because fatality rates differ by type of highway, six different types of highways were examined separately: rural Interstates, rural arterials, rural collectors, urban freeways, urban Interstates, and urban arterials. Separate regressions were fitted to data for 1981 and 1982; thus there were 12 separate regression models, 6 highway types for each of 2 years. Data were derived from published figures of the U.S. Department of Transportation (DOT), and the U.S. Department of Commerce. Data on state speeding citations are from unpublished compilations by the DOT. States with missing data were screened out: Alaska, Hawaii, and the District of Columbia were excluded from all regressions because of their markedly atypical highways and driving conditions. Table A-8 gives the characteristics of each subset of the data. The speed variance variable is a measure of the dispersion, or range, of speeds among drivers. This is approximately bell-shaped. The average speed is about at the center of this distribution, and the 85th percentile speed is about one standard deviation to the right. Thus, the 85th percentile speed less the average is a measure of the standard deviation of observed speeds.

TABLE A–8 Means of the Variables

Highway System	Fatality Rate[a]	Average Speed (mph)	Percent Drivers Exceeding 55 mph	Percent Drivers Exceeding 65 mph	85th Percentile Speed (mph)	Speed Variance (85–A)[b]	Citations per Driver[c]
1981							
Rural							
Interstate	1.81	58.1	68.7	9.31	63.2	5.13	74.0
Arterial	4.97	54.1	43.4	4.37	60.0	5.85	72.0
Collector	4.11	51.7	33.4	4.42	58.7	6.99	75.3
Urban							
Freeway	3.24	54.9	46.2	2.93	59.9	4.99	58.1
Interstate	1.37	55.8	54.1	4.46	61.1	5.33	59.4
Arterial	2.67	51.9	31.2	2.20	58.1	6.12	63.7
1982							
Rural							
Interstate	1.50	59.0	73.7	14.2	65.1	6.12	67.0
Arterial	4.24	54.4	47.1	6.23	61.2	6.81	64.8
Collector	4.32	51.8	35.4	5.54	59.6	7.85	67.3
Urban							
Freeway	1.77	56.2	55.1	6.80	62.5	6.32	53.9
Interstate	1.24	56.6	61.8	8.53	63.0	6.41	55.2
Arterial	2.33	52.2	35.9	4.02	59.4	7.25	55.7

[a]Fatalities per 100 million vehicle miles traveled.
[b]85th percentile speed minus the average speed: a rough measure of the standard deviation of the distribution of speeds.
[c]Speeding citations per 100 drivers per year.

TABLE A–9 Range of Variation Among States

Variable	Mean	Lowest Value	Highest Value
Urban Freeways			
Fatality rate	3.25	0.89	15.50
Average speed	54.90	51.10	57.20
Percent drivers exceeding 55 mph	46.30	23.10	64.60
Percent drivers exceeding 65 mph	2.94	0.40	8.10
85th percentile speed	59.90	57.40	63.80
85th percentile—average speed	5.00	1.80	7.50
Citations per driver	58.20	24.10	118.00
Rural Interstates			
Fatality rate	1.81	0.39	4.79
Average speed	58.10	54.80	62.50
Percent drivers exceeding 55 mph	68.70	40.30	88.80
Percent drivers exceeding 65 mph	9.31	1.70	28.60
85th percentile speed	63.20	58.80	69.40
85th percentile—average speed	5.13	2.70	9.10
Citations per driver	74.00	24.10	193.00

Considerable difference exists among states. The data in Table A-9 indicate this range of variation for two subsets of the data: Interstate rural roads in 1981 and urban freeways in 1981.

REGRESSION RESULTS

In simple models, it was difficult to find any significant relationship between highway speeds and highway fatalities; in complex models, the speed variables might become significant, but the signs were often perverse—negative instead of positive. Had the relationship been significant negative signs would indicate that higher speeds reduce fatalities. Other researchers who have conducted similar empirical studies indicated that they had encountered these same problems.

Independent Variables

The data in Table A-10 indicate the result from estimating a relatively simple model in the 12 subsets of the data (6 road types × 2 years). The model regressed fatality rate on average speed, while holding constant variables that measure medical accessibility and driver behavior in each state.

TABLE A–10 Effect of Speed on Fatality Rates

Highway Systems	*t*-Ratio of Average Speed
1981	
Rural	
Interstate	(0.4)
Arterial	(−2.3)
Collector	(−0.0)
Urban	
Freeway	(−1.6)
Interstate	(0.7)
Arterial	(−0.9)
1982	
Rural	
Interstate	(−0.7)
Arterial	(−0.9)
Collector	(−1.3)
Urban	
Freeway	(−0.5)
Interstate	(0.3)
Arterial	(−1.6)

Accessibility of Emergency Medical Care

The variable measuring accessibility to emergency medical care was quite difficult to formulate. The ideal variable would measure both the number of hospitals per square mile and the uniformity of their distribution. States with more hospitals per square mile offer better medical access to accident victims; but if all the hospitals are concentrated in a few areas, then someone having an accident in one of the areas without a hospital will not be able to receive fast medical attention. The variable created to measure medical access was hospitals per square mile multiplied by the proportion of population living in nonmetropolitan areas. This variable proved to be superior to all those examined.

Driver Behavior

After considerable investigation, discussed in the next section, it was determined that the best measure of driving behavior was speeding citations per driver. The data in Table A-10 indicate only the *t*-ratios and the significance levels for the speed variable. Only one of the *t*-ratios is significant, and nine of them have negative signs, which indicates that increased average speeds decrease the fatality rate. To understand the insignificance of the speed variables in these equations and the perverse signs of the speed variables in other equations, it is useful to construct a simple algebraic model of the effects of speed.

A Mathematical Model of the Fatality Rate

On a priori grounds, it appears reasonable that the fatality rate is a positive function of speed, a positive function of the variance in speeds, and a function of the effects of other variables (characteristics of road, driver, auto, etc.). This can be written as:

$$\begin{aligned} \text{Fatality Rate} &= F\,(\text{ speed, speed variance, other factors}) \\ &= a + b_1\text{Speed} + b_1 + b_2\text{Variance} + e \end{aligned}$$

where the other factors are temporarily put into e, the error term, to keep the algebra simple. Operationalizing speed as the average speed, and operationalizing speed variance as 85th percentile minus average speed, this initial equation translates into Equation 1:

$$\begin{aligned} \text{Fatality rate} &= a + b_1\text{Avg} + b_2(\text{85th Percentile} - \text{Avg}) + e \\ &= a + b_1\text{Avg} + b_2\ \text{85th Percentile} - b_2\text{Avg} + e \qquad (1) \\ &= a + b_2\ \text{85th Percentile} + (b_1 - b_2)\text{Avg} + e \\ &= a + b_2\text{85th Percentile} + b_3\text{Avg} + e \qquad (2) \end{aligned}$$

where b_3, the estimated regression coefficient in Equation 2, is actually the combined effect of speed and speed variance, $b_3 = b_1 - b_2$. If speed variance has more effect on the fatality rate than speed alone, that is, b_2 is larger than b_1, then b_3 will actually be negative. That is, it is expected that some estimated speed coefficients will be negative instead of positive, hence producing the apparently perverse regression coefficients observed in empirical research.

Table A-11 gives the result of estimating the model of Equation 2 on the 12 subsets of the data. As before, measure of hospital access and measure of driver characteristics are held constant. For each variable, the table gives the estimated regression coefficient to the left, and the *t*-ratio in parentheses beside it.

The data in the table confirm the algebraic model: 10 of 12 of the regression coefficients of average speed are negative, and 10 of 12 of the regression coefficients of 85th percentile speed are positive. (The R^2 figure given is corrected for degrees of freedom.)

TABLE A–11 Combined Effects of Speed and Speed Variance

Highway System	Average Speed	85th Percentile Speed	R^2
1981			
Rural			
Interstate	−0.24 (1.8)	0.20 (2.3)	0.62
Arterial	−0.75 (3.7)	0.58 (2.8)	0.25
Collector	0.007 (.0)	−0.01 (0.1)	0.00
Urban			
Freeway	−1.3 (1.7)	0.55 (0.7)	0.29
Interstate	−0.04 (0.3)	0.10 (1.2)	0.12
Arterial	−0.58 (2.4)	0.50 (2.2)	0.15
1982			
Rural			
Interstate	−0.21 (2.3)	0.19 (2.5)	0.52
Arterial	−0.41 (2.0)	0.35 (1.8)	0.08
Collector	−0.09 (0.7)	0.001 (.0)	0.10
Urban			
Freeway	−0.39 (0.8)	0.30 (0.7)	0.14
Interstate	0.04 (0.3)	−0.01 (0.1)	0.13
Arterial	−0.29 (1.7)	0.23 (1.2)	0.16

Examination of the size of the regression coefficients indicates that, in any given equation, both speed coefficients tend to be approximately equal to each other, though opposite in sign. But, by definition, $b_3 = b_1 - b_2$; hence if b_3 and b_2 are approximately equal, then b_1 must be near to zero. Because, from Equation 1, b_1 measures the effect of average speed on the fatality rate, this implies that speed has little or no effect on fatality rates.

It is easy to test this idea directly by rearranging the form in which the regressions are run. Accordingly, the model was reestimated in the form of Equation 1 because this yields direct coefficient estimates for the separate effects of speed and speed-variance. The results are given in Table A-12. The estimated coefficient for speed variance is positive in 11 of the 12 equations, and significant in 6 of these cases; it is insignificantly negative in one regression. The estimated magnitude and the *t*-ratios of average speed were insignificant in all 12 equations. Thus in this formulation speed, per se, had no effect on fatality rates; it was only variance that mattered.

A brief comment about the citations per driver variable is in order. This variable was added to the model to represent the effect of law enforcement on fatalities—states that enforce the speed limit strictly will issue more citations and this, in turn, should have a negative effect on the fatality rate. It is obvious, from Table A-12, that this is not true. Two other measures of law enforcement (policing expenditures per driver, number of police per driver) had the same negative coefficients. Because the number of tickets given is an interactive relationship between driver aggressiveness and police conscientiousness, states whose drivers speed frequently may decide to spend more money on law enforcement to bring speeding under control. Thus, it appears likely that citations per driver is primarily a measure of driver behavior. At least partial confirmation of this hypothesis can be observed in the negative correlation between the citation rate and average driver age: the higher the proportion of young, and presumably aggressive, drivers in a state, the higher the state's citation rate.

The paradoxical empirical results obtained earlier are explained by two comparisons. First, the insignificant speed coefficients arose in cases where speed alone was entered into the equation. Those coefficients were insignificant because speed alone does not have an effect. Second, the equations where some speed terms had negative coefficients, and some positive, result because the combination of speed terms enables the regression to control for variance effects. At least two of the percentile speed variables are required to represent variance algebraically: without at least two terms, none is significant; with at least two terms, thé coefficients become significant, and at least one is negative.

TABLE A–12 Final Regression Equations

Highway System	Speed Variance	Citations per Driver	Hospital Access	R^2	N
1981					
Rural					
Interstate	0.176 (2.3)	0.0136 (4.6)	−7.750 (3.6)	0.624	41
Arterial	0.677 (3.5)	0.0122 (1.6)	0.915 (0.2)	0.237	46
Collector	0.011 (0.1)	0.0041 (0.6)	−8.610 (1.6)	0.019	41
Urban					
Freeway	0.892 (1.3)	0.0634 (1.9)	−0.126 (1.1)	0.269	19
Interstate	0.103 (1.2)	0.0101 (2.0)	0.324 (0.2)	0.139	26
Arterial	0.526 (2.4)	−0.0187 (1.9)	−1.930 (0.5)	0.177	23
1982					
Rural					
Interstate	0.190 (2.6)	0.0071 (2.8)	−5.290 (3.7)	0.532	44
Arterial	0.375 (2.0)	0.0116 (1.7)	−0.424 (0.1)	0.101	47
Collector	0.046 (0.3)	0.0139 (2.4)	−0.830 (0.2)	0.089	41
Urban					
Freeway	0.281 (0.7)	0.0410 (2.5)	−2.860 (0.5)	0.193	18
Interstate	−0.011 (0.2)	0.0106 (2.8)	−0.168 (0.1)	0.167	27
Arterial	0.304 (1.8)	−0.0068 (1.2)	−5.720 (2.2)	0.168	21

Would raising the speed limit raise the variance and the fatality rate to their former levels? Perhaps it would; perhaps not. The answer depends on many aspects of driver behavior that are not well understood. Currently, however, the overall relationship between average speed and speed variance is generally negative today; the states with the highest average speeds tend to have the lowest variance. The data in Table A-13 indicate the correlations for the 12 subsets of the data. Further research is needed to determine if there are achievable ways to increase speeds without major impacts on speed variance and fatality rates (see Appendix H).

TABLE A–13 Correlation Between Average Speed and Speed Variance

Highway System	Correlation Coefficient Between Average Speed and Speed Variance
1981	
Rural	
Interstate	0.40
Arterial	−0.26
Collector	−0.15
Urban	
Freeway	−0.41
Interstate	0.00
Arterial	−0.17
1982	
Rural	
Interstate	−0.13
Arterial	−0.19
Collector	−0.27
Urban	
Freeway	0.16
Interstate	0.01
Arterial	−0.36

Appendix D
State Penalties For Violation of the 55 mph Speed Limit

State	Penalties for Exceeding Pre-55 mph Speed Limit	Penalties for Exceeding 55 mph Speed Limit
Alabama	$100 fine maximum, no points assessed.	*Conviction* / *Fine Maximum* / *And/or Jail Maximum* first / $5 to $100 / 10 days second / $200 / 20 days third / $500 / 6 months (within 1 yr) Points: 56 to 90 mph, 3; 91+ mph, 5.
Alaska	$200 fine maximum. *Note*: Until 1974 maximum speed was 50 mph.	$500 fine maximum and/or 90 days jail maximum. Points assessed.
Arizona	$100 fine maximum, up to 10 days imprisonment, 3 points.	*Conviction* / *Fine Maximum* / *And/or Jail Maximum* first / $100 / no jail second / $500 / 30 days third / $750 / 4 months Points, 3. Up to 65 mph $15 fine and no points and not applicable to insurance.
Arkansas	Within 1 year: first offense, $100 and/or 10 days; second offense, $200 and/or 20 days; third offense, $500 and/or 6 months. Points assessed.	Fine—same Points: 1 to 10 mph above speed limit, 3; 11 to 20 mph above speed limit, 4; 20+, 5; 100 mph or more, 6.

APPENDIX D (continued)

State	Penalties for Exceeding Pre-55 mph Speed Limit	Penalties for Exceeding 55 mph Speed Limit
California	Same as present.	Violations within one year: first conviction: $50 fine maximum, points; second conviction: $100 fine maximum, points; third conviction: $250 fine maximum, points.
Colorado	$25 fine maximum if not more than 15 mph above speed limit, $10 to $500 fine and/or up to 6 months imprisonment if more than 15 mph over limit.	1 to 9 mph above 55, 3 points, $100 fine maximum; 10 to 19 mph above 55, 4 points, $100 fine maximum; 20 mph above 55, 6 points, $300 fine maximum and/or 90 days imprisonment maximum.
Connecticut	$100 fine maximum, no points assessed.	Maximum fine $100, driver's license suspended for two convictions within 5 years for speeds above 70 mph. Violations of 55 mph now violation of the law, not an infraction.
Delaware	Within 2 years: first offense $25 to $200 fine and/or 10 to 30 days. Subsequent offenses $50 to $400 fine and/or 15 to 60 days.	1 to 5 mph above speed limit: $19.50 maximum; 6 to 15 mph above speed limit: $36.50 maximum; 16 to 20 mph above speed limit: $63.50 maximum; 21 mph and more above limit: $102.00 or more. Points: 56 to 61, 2; 62 to 65, 4; 66+, 5.
Florida	$500 fine maximum, up to 15 mph above limit 3 points assessed. In excess of 15 mph above limit 4 points assessed.	$500 fine maximum (average fine $25) Points: 56 to 70, 3; 71+, 4.
Georgia	$1,000 fine maximum, 12 months maximum imprisonment.	$1,000 fine maximum, and/or 12 months maximum imprisonment, points. (Average fine $40.00)
Hawaii	First offense $100 fine maximum; 10 days maximum imprisonment. Second offense $200 fine maximum; 20 days maximum imprisonment. Third offense $500 fine maximum; 6 months maximum imprisonment.	First offense, not more than $100; second offense within 1 year, not more than $200; third subsequent offense within 1 year, not more than $500. Points assessed.

Idaho	Up to $100 fine and/or up to 10 days imprisonment.	$5 fine maximum plus $7.50 costs, no points, nonmoving violation for insurance. This penalty applies to violations in excess of 55 mph but less than previously posted limits.
Illinois	First offense: maximum fine $500. Second offense: maximum fine $500. Third offense and subsequent offense $500 and/or 30 days in jail. Points assessed.	$100 fine maximum, 10 days jail maximum. Points assessed driver improvement action initiated with three moving violations. Speeding classed as a moving violation.
Indiana	First offense: maximum fine $100, 10 days maximum jail. Second offense: maximum fine $200, 20 days maximum jail. Third or subsequent offense: maximum fine $500, 6 months maximum jail.	$500 fine maximum (average fine $25.00) Points: 56 to 70 mph, 2; 71 to 80 mph, 3; 81+, 6.
Iowa	Up to $100 fine and/or up to 30 days imprisonment. Point system not used.	Maximum $100 fine 1 to 10 mph above limit: $10 fine; 11 to 20 mph above limit: $30 fine; 21 or more mph above limit, up to $100 and/or 30 days in jail; 1 to 6 mph above limit not recorded in driver record. Three violations in 12 months automatic suspension of license.
Kansas	Up to $500 fine.	Up to $500 fine, not entered on driver's record and not considered a moving violation unless exceeding 55 mph speed limit by 10 mph or more. No point system. Suspension for three or more moving violations committed on separate occasions.
Kentucky	$10 to $100 fine.	$10 to $100 fine maximum. No points for speed between 55 and 70 mph on Interstate and limited-access highways. Points assessed on all other 55 mph highways: 56 to 70, 3; 71 to 80, 6; 81+, 6 plus hearing and/or suspension up to 90 days.

APPENDIX D (continued)

State	Penalties for Exceeding Pre-55 mph Speed Limit	Penalties for Exceeding 55 mph Speed Limit
Louisiana	First offense, up to $100 fine and/or up to 30 days imprisonment. Second and subsequent, up to $500 fine and/or up to 90 days imprisonment.	First conviction: $100 fine, maximum and/or 30 days jail; 2 points assessed between 55 and 70 mph. Court cannot take license. Subsequent convictions: $500 fine maximum and/or 90 days jail.
Maine	Penalty pre-55 mph speed limit. When no other penalty is specifically provided fine shall not be less than $10 nor more than $100 or more than 90 days imprisonment or by both.	$10 to $100 fine. Average fine for 55 mph violation, $40.00. Points: 3 through 10, based on mph above the limit.
Maryland	Not more than $1,000 or not more than 6 months or both.	*Speed* / *Fine*: 56 to 64, $30; 65 to 74, $40; 75 to 84, $50; $500 fine maximum *Speed* / *Points*: 56 to 64, 1; 65 to 84, 2; 85 +, 5
Massachusetts	Up to $25 fine.	$50 fine. No point system, three moving violations within 1 year is cause for suspension (1983).
Michigan	Unless another penalty is provided fine shall not be more than $100 or more than 90 days or by both. Exceeding speed limit before 1974: by 10 mph or less, 2 points; by more than 10 and less than 15, 3 points; by more than 15 mph or by careless driving, 4 points.	$100 fine and/or 90 days jail maximum. No points for arrest for driving between 55 and 70 mph.
Minnesota	Imprisonment not more than 90 days or a fine of not more than $100. No point system.	$500 fine, maximum; 90 days jail, maximum; driver improvement action initiated by three minor misdemeanors. Speeding classified as minor misdemeanor.

Mississippi	Unless another penalty is provided fine is: first conviction, not more than $100 or more than 10 days; second conviction within 1 year, fine not more than $200 or more than 20 days or by both; third or subsequent conviction within 1 year after first conviction, fine not more than $500 or not more than 6 months or both.	First conviction: $100 fine, maximum and/or 10 day jail maximum. No point system. Three violations in 2 years, warning letter; 4 violations, interview and probation.
Missouri	Up to $500 fine and/or up to 1 year imprisonment. Points assessed.	$5 to $500 fine and/or up to 1 year imprisonment. No points assessed for speeds less than previous established speed limit. Speeding in excess of previous limit, 3 points assessed.
Montana	$10 to $100 fine and/or up to 10 days imprisonment.	Daytime: $5 fine, no points assessed. Nighttime: $12.50 to $100 fine, 3 points assessed.
Nebraska	Up to $500 fine and/or up to 6 months imprisonment.	Fine: 56 to 65, $10, no court costs or points (on Interstate system only). First offense not more than $100. Second offense not more than $200. Third offense not more than $300. Points same as 1974.
Nevada	$500 maximum fine, 6 months imprisonment maximum.	$10 up to 70 mph: no record, no points; above 70 mph, $500 maximum and points, possible jail.
New Hampshire	$100 maximum fine.	$100 fine maximum. No point system.
New Jersey	$200 fine maximum; 10 days jail maximum. Points assessed.	$200 fine maximum; 10 days jail maximum. Points assessed.
New Mexico	Up to 15 mph above speed limit, $5; 15 to 25 mph above speed limit $25; more than 25 mph above speed limit not misdemeanor, set by court.	Legislation now sets fine of $1 per mph above speed limit up to 70 mph. No points assessed. Motor Vehicle Division (MVD) cannot use conviction under this section for purpose of revocation or suspension of license unless speed was in excess of 70 mph. Above 70 mph, $200 fine maximum, 10 days in jail maximum. Points; 56 to 69, 2; 70 to 84, 4; 85+, 5. (Driver improvement action at 12 points).

APPENDIX D (continued)

State	Penalties for Exceeding Pre-55 mph Speed Limit	Penalties for Exceeding 55 mph Speed Limit
New York	Up to $100 fine and/or up to 30 days imprisonment.	Fine: first conviction, $100 and/or 30 days jail maximum; second conviction (within 18 months.) $200 and/or 90 days jail maximum; third conviction (within 18 months.) $300 and/or 180 days jail maximum. Points: 1 to 24 mph above speed limit, 3; 25 mph or more above speed limit, 5.
North Carolina	$100 fine, maximum and/or 60 days jail, maximum. Points assessed.	$100 fine, maximum and/or 60 days jail maximum. Points assessed.
North Dakota	Within 1 year: first offense $100 fine maximum or 10 days. Second offense $200 fine maximum or 20 days jail. Third offense $500 fine maximum or 6 months maximum.	*Speed (mph)* — *Fine* — *Points* 56 to 60 — $5 — 0 61 to 65 — $5 + $1 each mph above 60 — 1 66 to 70 — $10 + $1 each mph above 65 — 2 71 to 75 — $15 + $2 each mph above 70 — 3 76 to 80 — $25 + $3 each mph above 75 — 4 81 to 90 — $40 + $3 each mph above 80 — 6 91 to 100 — $70 + $3 each mph above 90 — 8 101+ — $100 + $5 each mph above 100 — 12
Ohio	$250 maximum fine, 30 days maximum.	$100 fine. No points assessed for driving between 55 and 70 mph. Two points assessed for speeds in excess of 70 mph.
Oklahoma	$10 to $200 fine and/or 5 to 30 days.	Same fine and imprisonment. Cannot record violation in driver record or cancel insurance for exceeding 55 mph. No points assessed unless above previous speed limit.

Oregon	$500 fine maximum. Up to 1 year imprisonment.	Maximum fine $100. Cannot suspend or revoke license for accumulation of above 55 mph speeding violations only.
Pennsylvania	$10 fine regardless of speed plus $5 court cost; total penalty $15.	Fine: $35 plus an addition of $2.00 per mph in excess of 5 mph above speed limit, plus $10 court costs. *Points Assessed* *Speed Over Limit* — *Points* 6 to 10 — 2 11 to 15 — 3 16 to 25 — 4 26 to 30 — 5 31 and above — 5+ Departmental hearing resulting in possible: (i) driver improvement school and/or (ii) special examination and/or (iii) 15-day suspension.
Rhode Island	Unless otherwise noted, not more than $500 or more than 1 year or both.	$15 fine, 1 to 5 mph above 55; $20 fine, 6 to 10 mph above 55; $25 fine, 11 to 15 mph above 55. No point system.
South Carolina	Less than 10 mph above posted limit, not less than $5 nor more than $25 or more than 10 days imprisonment. More than 10 mph but less than 25 mph above posted limit, not less than $10 nor more than $50 or more than 20 days imprisonment. More than 25 mph above posted limit, not less than $25 nor more than $100 or not less than 10 nor more than 30 days imprisonment.	$100 fine maximum, 10 mph or less above 55. Above 55 mph, 2 points; 10 mph above 55 mph, 4 points.

APPENDIX D (continued)

State	Penalties for Exceeding Pre-55 mph Speed Limit	Penalties for Exceeding 55 mph Speed Limit
South Dakota	Up to $100 fine.	Fine: $2.00 per mile above 55 mph plus court costs, $100 maximum. *Interstate Highway* / *Points*: 56 to 60, 1; 61 to 75, 2; 76+, 3 *Other Highway* / *Points*: 56 to 60, 1; 61 to 65, 2; 66+, 3 Suspension at 15 points in 12-month period, 22 points in 24-month period. No access to driver record for violations less than 70 mph.
Tennessee	$2 to $50 fine/and or up to 30 days imprisonment plus $2 moving violation.	Between 55 and 75 mph, $2 to $50 fine and/or up to 30 days jail. No court costs and not entered in driver record. Reckless driving may not be charged.
Texas	$1 to $200 fine plus court costs.	$1 to $200 fine plus court costs. No point system. Driver license suspension based on hazardous violations: four in 1 year or seven in 2 years.
Utah	Fine not to exceed $299 (Class B misdemeanor).	$300 fine maximum; 6 months jail maximum Points: 56 to 64 mph, 35; 65 to 74 mph, 55; 75+ mph, 75.
Vermont	Up to $100 fine.	$2.00 per mile above 55 mph. Points: 1 to 19 mph above speed limit, 3; 20+ mph above speed limit, 6.
Virginia	$10 to $100 fine and/or 1 to 10 days imprisonment.	Fine: 56 to 74 mph, $18.00 + $2.00 mph above 55 mph; 75+ mph—violator charged with reckless driving. Points: 56 to 64 mph, 2; 65 to 74 mph, 4; 75+ mph, 6.

Washington	Maximum fine $250; Maximum jail term, 90 days.	$250 fine maximum; 90 days jail maximum. No points.
West Virginia	First offense: maximum fine $100 not more than 10 days in jail. Second offense: maximum fine $200 not more than 20 days in jail. Third offense: maximum fine $500 not more than 6 months in jail.	Fine: $100 to $500 maximum. Points: 56 to 74, 3; 20+ mph, 6. Accumulation of: 9 points, hearing and/or suspension; 12 points revocation for not more than 1 year.
Wisconsin	Fine $20 to $200; not more than 30 days in jail. Points, plus automatic 15-day suspension for 20+ mph above speed limit.	Same, except automatic suspension at 25+ above limit.
Wyoming	Up to $100 fine and/or up to 6 months imprisonment.	$100 fine maximum. No point system. Three moving violations within 1 year may result in hearing and suspension. Violations between 56 and 75 mph do not count.
Puerto Rico		$50 fine maximum and/or 10 days jail maximum. Points: From three to seven based on mph above limit.

SOURCE: *55 mph Fact Book*. NHTSA, U.S. Department of Transportation, 1979. Updated information provided by NHTSA.

Appendix E
Analysis of Speedometer Adjustments

The analysis of speedometer adjustments in this appendix is provided by John J. Wiorkowski, Professor, Programs in Mathematical Science, University of Texas, Dallas. The adjustments for speedometer error as described in the October 21, 1982, Memorandum to the FHWA regional administrators[1] lack logical justification. The issue, however, is complicated, as described in the following analysis.

Basically, data are collected on vehicle speed (for example, x). The variable of real interest, however, is the speed that the driver intends to drive at (for example, y). These two numbers are not the same due to speedometer error (for example, u), but are related through the relationship

$$y = x + u$$

That is, the intended speed is the actual speed as measured with speedometer error. Under a 55 mph speed limit, a driver intends to speed if $y \geqslant 55$ mph, however, whether or not the driver actually speeds depends on his speedometer. Suppose a driver travels at 57 mph. His speed really could be 54 mph with a positive speedometer error of 3 or 60 mph and a negative speedometer error of -3 mph. If an intentional speeder is one whose speedometer registers higher than 55 mph, then compliance should be measured by estimating Pr $(y > 55)$. However, data are collected only on x, the actual speed, and information is available on speedometer error distributions. The goal then is to estimate Pr $(y > 55)$ given information on x and u.

Assume that the counting device can measure the proportion of cars traveling at speed x, and denote this by f(x). Then conceptually f(x) is defined on all speeds from 0 to some maximum, and

$$\sum_{x} f(x) = 1$$

Assume further that the range of absolute speedometer error is ± 5 mph and that the proportion in error by an amount of u is given by g(u) so that

$$\sum_{u=-5}^{5} g(u) = 1$$

Then assuming that x and u are independent (i.e. that speedometer error is unrelated to speed, an assumption that should be approximately valid for high speeds), one can determine the probability distribution of y, the speedometer speed. Specifically,

$$\Pr(y = y_0) = \sum_{u=-5}^{5} f(y_0 - u)g(u) \tag{1}$$

In expanded form,

$$\begin{aligned} \Pr(y = 56) = {} & f(61)g(-5) + f(60)g(-4) + f(59)g(-3) \\ & + f(58)g(-2) + f(57)g(-1) + f(56)g(0) \\ & + f(55)g(1) + f(54)g(2) + f(53)g(3) \\ & + f(52)g(4) + f(51)g(5) \end{aligned}$$

It follows from Equation 1 that

$$\Pr(y > 55) = \sum_{y=56}^{\text{max speed}} \sum_{u=-5}^{5} f(y-u)g(u) \tag{2}$$

Focusing on the 55 mph speed limit, Equation 2 can be rewritten as

$$\Pr(y > 55) = \sum_{s=1}^{\text{max speed}-55} \sum_{u=-5}^{5} f(55+s-u)g(u) \tag{3}$$

Now the probability that x > 55 can be written as

$$\Pr(x > 55) = \sum_{k=1}^{\text{max speed}-55} f(55+k) \tag{4}$$

Thus to determine over or underutilization, it is necessary to compare Equations 3 and 4. To make this comparison, it is necessary to rewrite

Equation 3 to sum over k. To do this, specific speeds must be examined. If vehicles traveling at 51 mph are examined, then only if the speedometer is in error by +5 mph will they appear to be speeding. For vehicles traveling 52 mph, the speedometer will read greater than 55 mph if there is an error of +4 mph or +5 mph. The actual speeds and the errors are summarized in Table A-14.

For speeds of 61 mph or greater the contribution will be, for example,

$$f(61) \sum_{u=-5}^{5} g(u), \text{ but } \sum_{u=-5}^{5} g(u) = 1$$

Therefore any speed x above 60 mph will contribute a full f(x). Combining these facts, Equation 3 can be rewritten as

$$\Pr(y > 55) = \sum_{k=-4}^{5} f(55+k)\left(\sum_{u=1-k}^{5} g(u)\right) + \sum_{k=6}^{(\max-55)} f(55+k) \quad (5)$$

TABLE A–14 Summary of Actual Speeds and Speedometer Errors

Actual Speed	Errors	Contribution to Pr(y ≥ 55)
51	5	f(51)g(5)
52	4,5	f(52)[g(4) + g(5)]
53	3,4,5	f(53)[g(3) + g(4) + g(5)]
54	2,3,4,5	f(54)[g(2) + g(3) + g(4) + g(5)]
55	1,2,3,4,5	$f(5)\left[\sum_{u=1}^{5} g(u)\right]$
56	0,1,2,3,4,5	$f(56)\sum_{u=0}^{5} g(u)$
57	−1,0,1,2,3,4,5	$f(57)\sum_{u=-1}^{5} g(u)$
58	−2,−1,0........,5	$f(58)\sum_{u=-2}^{5} g(u)$
59	−3,−2,−1,.........,5	$f(59)\left(\sum_{u=-3}^{5} g(u)\right)$
60	−4,−3,−2,−1,.........,5	$f(60)\left(\sum_{u=-4}^{5} g(u)\right)$

Now Equation 4 can be rewritten as

$$\Pr(x > 55) = \sum_{k=1}^{5} f(55+k) + \sum_{k=6}^{(\max - 55)} f(55+u) \qquad (6)$$

The last term of Equations 5 and 6 is the same (this term is simply the Pr(x > 60 mph). Thus subtracting Equation 6 from Equation 5 yields this result

$$\Pr(y > 55) - \Pr(x > 55) \qquad (7)$$

$$= \sum_{k=-4}^{5} f(55+k) \left(\sum_{u=1-k}^{5} g(u) \right) - \sum_{k=1}^{5} f(55+k)$$

Equation 7 indicates that an adjustment may indeed be necessary, but the magnitude of the adjustment may be small. Further, it is unclear whether the adjustment is negative and indeed under some circumstances it may be positive. To better understand this adjustment, examine Table A-15, which is the right side of Equation 7 written term by term. The g values with negative entries result from

$$\sum_{u} g(u) = 1$$

Thus from Equation 7 the multiplier of f(56) is $g(0) + g(1) + g(2) + g(3) + g(4) + g(5) - 1$, but because all the g's add to 1, the net effect is $-[g(-1) + g(-2) + g-(3) + g(-4) + g(-5)]$. Now if g(u) is a symmetric error distribution (e.g., if g(u) were normal or uniform), then $g(-u) = g(u)$ so that the right side of Equation 7 (the correction) could

TABLE A–15 Right Side Of Equation 7

Speed	Probability	Multiplier
51	f(51)	g(5)
52	f(52)	g(4) + g(5)
53	f(53)	g(3) + g(4) + g(5)
54	f(54)	g(2) + g(3) + g(4) + g(5)
55	f(55)	g(1) + g(2) +g (3) +g (4) + g(5)
56	f(56)	−[g(−1) + g(−2) + g(−3) + g(−4) + g(−5)]
57	f(57)	−[g(−2) + g(−3) + g(−4) + g(−5)]
58	f(58)	−[g(−3) + g(−4) + g(−5)]
59	f(59)	−[g(−4) + g(−5)]
60	f(60)	−g(−5)

be written as

$$\begin{aligned}\Pr(y > 55) - \Pr(x > 55) &= [f(51) - f(60)]g(5) \\ &+ [f(52) - f(59)][g(4) + g(5] \\ &+ [f(53) - f(58)][g(3) + g(4) + g(5)] \\ &+ [f(54) - f(57)][g(2) + g(3) + g(4) + g(5)] \\ &+ [f(55) - f(56)][g(1) + g(2) + g(3) + g(4) + g(5)]\end{aligned} \tag{8}$$

From Equation 8 it follows that if the actual speed distribution is symmetric about 55.5 mph, then no correction is necessary.

Further, if the actual speed distribution is shifted to the right of 55.5 mph, then the correction will be negative; but if the actual speed distribution is shifted below 55.5 mph, then speedometer error requires a positive correction.

Equation 8 also shows the error of the method outlined in the December 8, 1980, memorandum to the regional administrators of NHTSA and FHWA.[2] Assuming that $g(x) = 0.1$ (which corresponds to the assumption that $g(x)$ is uniform on ± 5 mph but again incorrectly because in that case $g(x)$ would be 1/11 not 1/10), the correction would be

$$\begin{aligned}&-f(60)g(5) - f(59)[g(4) + g(5)] \\ &-f(58)[g(3) + g(4) + g(5)] \\ &-f(57)[g(2) + g(3) + g(4) + g(5)] \\ &-f(56)[g(1) + g(2) + g(2) + g(4) + g(5)]\end{aligned} \tag{9}$$

which as observed by comparison with Equation 8 is only the negative part of the correction.

Finally, the assumption that $g(u)$ is uniform is extremely surprising. Although empirical studies will possibly show that $g(u)$ is symmetric, it is expected that it would have been approximately normal. Thus, this assumption on the part of NHTSA and FHWA administrators compromises their methodology. Further, they assume the distribution in the range $56 \leqslant x \leqslant 60$ is uniform; that is, that $f(56) = f(57) = f(58) = f(59) = f(60)$. It would be surprising if this were true. Thus, their adjustment methodology is flawed and probably overestimates the adjustment, if an adjustment is necessary.

What would be computed with adjustment in Equation 9 can be shown to be the joint probability that $x > 55$ and $y > 55$; i.e., the proportion of cars exceeding the speed limit and whose drivers intend to exceed the speed limit. Thus drivers who think they are speeding, but because of faulty speedometers are in reality not speeding, are not counted. This might seem reasonable except that the NHTSA-FHWA memorandum states, "The assumption is that a driver who knowingly exceeds the speed limit should be included in the 'percent exceeding' statistic calculation." Thus, individuals are incorrectly excluded.

NOTES

1. The adjustment for speedometer variability, speed measuring error, and statistical error as stated in a memorandum from the Director of the Office of Traffic Operations, Federal Highway Administration, to Regional Federal Highway Administrators, October 21, 1982 are listed below.

   The allowable adjustment methods for the fiscal 1982 certifications are as follows:

   1. *Speedometer Variability*

      Let A = percent exceeding 55 mph from the speed monitoring program,
      B = percent exceeding 60 mph,
      C = (A − B),
      D = adjusted percent exceeding 55 mph, and
      $D = 0.7 \times C + B$

   2. *Statistical Error*

      For the percent exceeding 55 mph, use as the work value the lower bound of the confidence interval based on a one-tailed test.
      Let D = percent exceeding 55 mph adjusted for speedometer variability,
      E = percent exceeding 55 mph adjusted for speedometer variability and statistical error, and
      E = D − 1.645 × (the standard error of the statewide percent exceeding 55 mph).

   3. *Speed Measuring Equipment Error*

      Any additional error factor that a state can determine and document.

2. The following text is quoted from the December 8, 1980, memorandum from the Administrators of NHTSA and FHWA to the Regional Administrators of these agencies.

Due to limited data available on speedometer error distributions and frequency distributions of vehicle speed, any methodology for calculating speedometer errors must be based upon a set of reasonable assumptions. The following illustrates one methodology accompanied by a given set of assumptions.

Assumptions

1. Speedometer error ranges from +5 mph to −5 mph.
2. There are equally as many negative errors as positive errors or the probability of a negative error is 0.5.
3. The distribution of the negative errors is uniform rather than unimodal.
4. The frequency distribution of vehicles traveling between 56 and 60 mph is uniform.

A negative speedometer error occurs when the actual traveling speed is greater than the speed indicated by the vehicle's speedometer. For example, if a vehicle is actually traveling at 60 mph and the speed indicated by the speedometer is 56 mph, a negative speedometer error of 5 mph has occurred.

Positive speedometer errors need not be adjusted because vehicles exceeding 55 mph will indicate speedometer speeds higher than actual travel speeds. The assumption is that a driver who knowingly exceeds the speed limit should be included in the "percent exceeding" statistic calculation.

The speed range requiring adjustment, therefore, is limited to the maximum *negative* speedometer error. For example, if a vehicle is traveling at the actual speed of 65 mph and the speedometer error is 5 mph negative error, the speedometer is reading 61 mph which is in violation of the 55 mph speed limit. Therefore, the speed range requiring adjustment for speedometer error is simply the maximum negative error added to 55 mph. In this case the range to be adjusted will be speeds exceeding 55 mph but not exceeding 60 mph.

Since it is not known how many vehicles are traveling at 55, 56, 57, 58, 59, and 60, nor how many vehicles have negative error speedometers of 1, 2, 3, 4, and 5 mph, assumptions 3 and 4 have been made; i.e., the numbers of vehicles traveling at each mile interval from 55 through 60 mph are equal and half of the vehicles within each mile interval have negative speedometer errors equally distributed from 1 to 5 mph.

To calculate the adjustment due to speedometer errors, assume that half of the vehicles traveling at 56 mph have a negative speedometer error of at least one mile per hour. Therefore one half of the percent of vehicles traveling at 56 mph should be eliminated from the "percent exceeding" statistical calculation. For vehicles traveling 57 mph, one half of them have a negative error and it is assumed that 80 percent have negative errors exceeding one mile per hour. Therefore, the adjustment to be made for vehicles traveling at 57 mph is $.8 \times .5 = .4$. If this process is continued through the 56 through 60 mph speed range, then the adjustments are:

0.5(56 mph freq.) + 0.4(57 mph freq.) + 0.3(58 mph freq.) + 0.2(59 mph freq.) + 0.1(60 mph freq.)

Since all the frequencies are assumed equal, then the adjustment to each mile across the speed range is $(0.5 + 0.4 + 0.3 + 0.2 + 0.1) \times$ (percent of vehicles exceeding 55 mph for each mile in the speed range) $= 1.5 \times$ percent of vehicles exceeding 55 mph for each mile in the speed range will yield the appropriate adjustment for speedometer error.

Appendix F
Survey of State Traffic Enforcement Administrators

To supplement available national data on speed-limit enforcement efforts, the state agencies responsible for enforcement of the 55 mph speed limit were surveyed. Thirty-eight states responded, although not all surveys contained complete information. Few states have consistent information on key measures of enforcement efforts, which limits the ability to systematically compare states. Nonetheless, the responses provide useful information about general trends in enforcement, the attitudes of police administrators toward the 55 mph speed limit, and point to some of the inherent limits of enforcement.

ALLOCATION OF PERSONNEL

State patrol and state police agencies are responsible for speed-limit enforcement, crime and accident investigation, enforcement of drunken driving crackdowns, and many other similar tasks. Some of these responsibilities occur on highways other than those posted at 55 mph. Overall, states allocate about 20 percent of patrol staff hours to highways posted below 55 mph. The remaining time (80 percent) is spent on highways posted at 55 mph, but not all of this time is spent simply enforcing the 55 mph speed limit. Most agencies expect their officers to enforce all relevant traffic regulations while patroling the highway; thus estimates of the time spent enforcing only the 55 mph speed limit are not available. Indeed, some states could not provide estimates of the amount of time spent patroling different types of highways.

The majority of state patrol staff hours are deployed on rural highways. Patrol staff hours classified by highway system are given in Table A-16. These highways comprise the majority of mileage posted at 55 mph. This allocation, however, is not entirely optimal either from the standpoint of total travel on these highways or total motor vehicle deaths. For example, 29 percent of patrol staff hours are devoted to the rural Interstates, which carry 19 percent of total travel, but account for only 9 percent of total fatalities. In contrast 57 percent of patrol staff hours are devoted to other rural highways, which carry 44 percent of travel on highways posted at 55 mph, and which account for 72 percent of fatalities. A comparison of staff-hour allocations with travel and motor vehicle fatalities on highways posted at 55 mph is given in Table A-17.

EFFICIENCIES IN DEPLOYMENT

The discrepancies between deployment, travel, and safety are partly explained by limited enforcement resources. The most concentrated travel occurs on the rural Interstate. Although this system only accounts for 6 percent of mileage it carries 19 percent of traffic. As a result limited enforcement resources can be deployed where most travel occurs. If state agencies were to deploy all their officers on rural arterials the number of miles to be covered would increase from 31,500 miles of rural Interstates to more than 200,000 miles of rural arterials. Patrol officers would be spread much more thinly and would have contact with far fewer travelers.

TABLE A–16 Patrol Staff Hours Classified by Highway System (N = 26)

Highway System	Number	Percent
Urban		
Interstate	826,021	7.1
Other multilane highways	512,805	4.4
Arterials	192,777	1.7
Rural		
Interstate	3,401,975	29.4
Other multilane highways	2,843,786	24.5
Rural two-lane	3,812,272	32.9
Total	11,589,636	100.0

TABLE A–17 Comparison of Staff-Hour Allocations with Travel and Motor Vehicle Fatalities on Highways Posted at 55 mph

Highway System	Manhours (%)	Travel (%)	Fatalities (%)
Rural			
Interstate	29.4	18.9	9.1
Other 55 mph highways	57.4	44.3	72.4
Urban			
Interstate	7.1	21.2	7.6
Other 55 mph highways	6.1	15.6	10.8
Total	100.0	100.0	100.0

Shared Enforcement

Another reason that state patrols allocate most of their time to main rural highways is explained by the shared administrative responsibilities for enforcement. In many states enforcement of speed limits is shared on highways that traverse urban county and municipal jurisdictions (Table A-18). Although the state agencies often have primary responsibility for enforcing 55 mph speed limits, even when these highways cross jurisdictional lines, in some states primary responsibility is transferred to local governments. Less than one-third of state patrols responding to the survey have sole responsibility for enforcing the 55 mph speed limit on county and city routes.

Some respondents to the survey indicated problems with shared speed-limit enforcement. In Michigan, for example, the state's most populous urban country shares responsibility for enforcing the speed limit but does not have active patrols for enforcement of the 55 mph speed limit. One state administrator indicated that in his state the state patrol had

TABLE A–18 States with 55 mph Speed Limit Enforcement Shared Among State, County, and City by Type of Highway (N = 38)

Highway System	State Only (%)	State and Local (%)	Not Applicable (%)
Interstates	28.9	71.1	–
State highways	23.6	76.3	–
County roads	31.6	60.5	7.9
City roads	21.1	68.4	10.5
Other	15.8	28.9	55.3

only 13 percent of police officers but issued 75 percent of all speeding citations for violations of the 55 mph speed limit. He indicated that speed-limit enforcement was not a priority for local agencies. The shared enforcement on urban highways, and the lower priority given to speed-limit enforcement by local governments, may help to explain why urban highways have demonstrated the greatest increases in noncompliance with the 55 mph speed limit in recent years.

Trends in State Enforcement of the 55 mph Speed Limit

Among the 38 states that responded to the survey, enforcement of the 55 mph speed limit has remained fairly constant. The number of patrol officers and the number of citations issued have increased slightly since 1977 as indicated below.

Citations for Exceeding 55 mph	1977	1980	1983	Percentage Change 1977–1983
(N = 20) (000s)	4,053	4,401	4,204	+3.7

Patrol Officers	1975	1977	1980	1983	Percentage Change 1975–1983
(N = 20)	12,241	12,106	12,317	12,311	+0.5

A small decline in the number of citations issued occurred between 1980 and 1983. This is partly explained by the increased use of written warnings in many states. Written warnings serve to remind motorists of the speed limit and its importance without penalizing them. The Ohio State Highway Patrol promotes the use of written warnings for a variety of reasons. A written warning, even without a penalty attached, has a greater effect on motorist behavior than a verbal admonition to slow down. The warnings issued by the Ohio State Patrol include a reminder of the importance of slower travel to safety. In addition, when other motorists observe speeders being stopped, the effect of the written warning is compounded. One-half of the states responding to the survey indicated that they have begun using written warnings as an enforcement tool in recent years, and the number of written warnings issued increased 36.5 percent between 1980 and 1983.

Most state agencies are currently under hiring freezes and state patrols are no exception. Some states rely heavily on the use of overtime pay for patrolmen to increase speed-limit enforcement, which allows the agency to increase the number of patrols without hiring additional personnel. One state official indicated that 20 percent of 55 mph enforcement is achieved through overtime pay. The average state, however, achieves less than 5 percent of 55 enforcement through overtime pay.

Police Chief Opinions on Speed Limit Enforcement

The majority of the 38 police chiefs surveyed (60.5 percent) believes that the 55 mph speed limit should be retained. The survey results are as follows:

Support (%)	Oppose (%)	No Opinion (%)
60.5	26.3	13.2

Those opposed (26.3 percent) noted in written comments that the high level of violations on rural Interstates indicates that the speed limit should be modified to coincide with motorist behavior and complained of the difficulty of limiting speeds to 55 mph on highways designed for higher speeds.

The majority of the 38 chiefs surveyed also believes that state traffic courts are strict enough on speeders in their states, as indicated by the following survey results:

Strict Enough (%)	Not Strict Enough (%)	No Opinion (%)
60.5	34.2	5.3

In the small number of states that provided data on conviction rates for speeders cited for violating the speed limit, conviction rates were well above 90 percent. Part of the explanation for these high conviction rates is the speed at which citations are issued. In the few states that provided data on the speeds at which citations were issued, the majority of citations was issued for speeding violations in the 65 mph plus range.

Most patrol agencies provide some tolerance in speed-limit enforcement. More than one-half of the states responding to the survey have formal policies, written or unwritten, that indicate the amount of tolerance that officers should permit before writing a citation. These tolerances are given in the table below. Twenty-six percent permit a tolerance of up to 5 mph, 21 percent allow a tolerance of between 6 and 10 mph, and 5 percent allow speeds 10 mph above the posted limit.

No Tolerance Policy (%)	0–5 mph (%)	6–10 mph (%)	Over 10 mph (%)	Officer Discretion (%)
23.7	26.3	21.0	5.3	23.7

Preferred Strategies for Increasing Compliance

The majority of police chiefs believes that the best technique for improving compliance with the speed limit would be to increase the number of officers patroling the highways. Alternative strategies for increasing compliance are given in Table A-19. Currently for every on-duty state patrolman there are 190 miles of highway posted at 55 mph. Of necessity, therefore, many miles of highway currently remain unpatroled.

An increase in the number of citations issued to speeders is ranked as the second most effective stategy for improving compliance. The third most effective policy is less clear-cut; the police chiefs were almost evenly divided among increased numbers of citations, increased penalties, and greater reliance on public information campaigns to educate motorists about the importance of compliance. Increased penalties emerged as a majority choice, but only among the least effective strategies as indicated in Table A-19.

When asked how they would allocate an additional $1 million to their budgets if the goal was simply to improve safety on state highways, the

TABLE A–19 Ranking of Alternative Strategies for Increasing Compliance (N = 38)

Strategy	First (%)	Second (%)	Third (%)	Fourth (%)	Fifth (%)	Total (%)
Increased patrols	73.7	10.6	13.2	0.0	2.6	100.0
Increased citations	0.0	42.1	28.9	26.3	2.6	100.0
Increased penalties	10.5	15.8	21.0	36.8	15.8	100.0
Public information campaigns	10.5	28.9	26.3	23.7	10.5	100.0
Other	5.3	2.6	10.5	13.2	68.4	100.0
Total	100.0	100.0	100.0	100.0	100.0	

police chiefs allocated 29 percent to additional staff hours for enforcement of the 55 mph speed limit and 12.6 percent to capital budgets (mostly for additional speed-monitoring equipment and vehicles). A description of resource allocations to increase highway safety is given in Table A-20. The individual choice that received the most increased funding to improve safety was to crack down on drunken driving.

CONCLUSION

The results of this survey indicate some of the limitations inherent in enforcing the 55 mph speed limit. State agencies tend to allocate resources for their enforcement efforts on rural Interstates to increase patrol efficiency. As a result less effort is devoted to other rural highways. These routes, however, tend to be far less safe. Speed-limit violations on two-lane rural roads pose a greater risk to safety than speed-limit violations on highways of Interstate quality. However, the enormous number of miles of highway posted at 55 mph cannot be equally covered. The low priority assigned to speed-limit enforcement in some urban areas by local governments may help to explain why the data for these highways indicate the greatest increase in noncompliance in the last 2 years.

Overall it appears that most state enforcement administrators support the 55 mph speed limit. Enforcement efforts, measured by patrol officers deployed and citations issued, have remained fairly constant in recent years. Most police chiefs view an increase in patrol staff hours as the most effective approach for improving motorist compliance with the law. Only a minority believe that increasing the penalties for speed violations is an effective policy for improving compliance.

TABLE A–20 Resource Allocations to Increase Highway Safety (N = 37)

Budget Item	Total ($)	Percent
Increased staff hours for 55 mph enforcement	10,740,000	29.1
Purchase equipment: radar, vehicles	4,670,000	12.6
Crackdown on drunken driving	13,580,000	36.7
Public information campaigns	3,010,000	8.1
Other	5,000,000	13.5
Total	37,000,000	100.0

Appendix G
Estimates of Taxpayer Benefits

OLD AGE, SURVIVORS, AND DISABILITY INSURANCE

Total Old Age, Survivors, and Disability Insurance (OASDI) costs come from three separate programs: lump sum awards, survivors benefits, and disability insurance. The savings attributed to the 55 mph speed limit totaled $40 million (in 1983 dollars). Calculations for each program follow.

Total OASDI = Lump Sum + Survivor Benefits + Disability Benefits

$39,588,947 = 779,736 + 37,465,809 + 1,343,397

Lump Sum Awards

[Estimated Lives Saved] × [Proportion by Sex]
0.73 Male
0.27 Female
SOURCE: (*1*)

× [Proportion over 15]
0.935 Male
0.894 Female
SOURCE: (*1*)

× [Proportion Adults Eligible]
0.93 Male
0.72 Female
SOURCE: (*2*)

× [Average Lump Sum Award]
$253.00 (1980 dollars)
SOURCE: (*3*)

× [Administrative Cost]
3.3 percent of total
SOURCE: (*4*)

Calculation

[3000 × (0.73)(0.935)(0.93)(253)] = 481,792
[3000 × (0.27)(0.894)(0.72)(253)] = 131,909
+ 613,701 (1.033) = 633,953
Administrative costs (1980 dollars)

\$633,953 × CPI = \$779,736 (1983 dollars)
(1.23)

Survivor Benefits

[Estimated Lives Saved] × [Proportion by Sex] × [Proportion over 15]
(same as above) (same as above) (same as above)
× [Proportion Eligible] × [Proportion of victims who have surviving family members]

0.73
SOURCE: (*5*)

× [Average Annual Benefit] = First Year Costs

\$3,360
SOURCE: (*6*)

Total Costs = [First Year Cost] × [Average Length of Payments]

7.2 years
SOURCE: (*5*)

Discounted Total Cost = [First Year Cost] × [Discount factor for 7.2 yr]

4.956 assuming a discount rate of 10 percent

Calculation

[3,000 × (0.73)(0.935)(0.93)] = 1,904
+ [3,000 × (0.27)(0.894)(0.72)] = 521
= 2,425 (0.73)(3360) = \$5,949,745

Total Cost = 6,941,369 (7.2)(1.033)(1.23) = \$54,429,747

Discounted = 6,941,369 (1.033)(4.956)(1.23) = \$37,465,809

Disability Benefits

[Number AIS 4 and 5 injuries averted] × [Proportion by Sex]
(assume same as for fatalities)
× [Proportion over 15] × [Eligibility] × [Average Benefit]
(same as above) (same as above) $2,334
SOURCE: (7)
× [Administrative Cost] = Total (1980 dollars)
Total × 1.23 = Total in 1983 dollars

Example:

[560(0.73)(0.93)(0.935)]
+ [560(0.27)(0.72)(0.895)]
= 453 (2,334)(1.033)(1.23) = $1,343,397

ESTIMATES FOR MEDICAID AND MEDICARE

The cost of the Medicaid and Medicare programs are estimated in a similar way. Both depend on assumptions about the number and type of injury. In this estimate the injuries averted that can be attributed to the 55 mph speed limit are taken from the estimates derived in Chapter 4. The mid-points for AIS 1, 2, and 3 injury estimates are used, and the estimates for injuries in the AIS 4 and 5 categories are taken from the high end of the range. As noted in that discussion, because of the estimating techniques used these estimates are more accurate.

The first estimate is derived in the following way:

Injury Category	Number Injuries Averted	Average Cost ($)	Total Cost Savings (000s)
AIS 1	39,000	223	8,697
AIS 2	8,500	1,850	15,725
AIS 3	2,900	4,233	19,472
AIS 4	400	12,696	5,078
AIS 5	160	106,990	17,118
			66,090

SOURCE: (*8*)

This weighted cost estimate is then multiplied by the appropriate Medicaid and Medicare adjustment factor to derive the estimate of the savings attributed to the 55 mph speed limit.

Medicaid

Savings = \$66,090,000 × [Proportion population receiving Medicaid]
0.098
SOURCE: (*9*)
× [Administrative adjustment] = \$6,787,707
1.048

Medicare

Savings = \$66,090,000 × [Eligibility] × [Proportion Elderly Injured]
0.95 — SOURCE: (*10*)
0.107 (same as fatality) — SOURCE: (*1*)
×[Proportion of cost paid by general fund]×[Administrative Cost]
0.6834 — SOURCE: (*11*)
1.057 — SOURCE: (*11*)

Calculation

\$4,852,808 = 66,090,000 (0.95) (0.107) (0.6834) (1.057)

SUPPLEMENTAL SECURITY INCOME (SSI)

Savings = [Number AIS 3–5 Injured] × [Proportion of Population Receiving SSI]
0.018
SOURCE: (*12*)
× [Average Benefit] × [Administrative Adjustment] × [CPI]
\$2,196 — SOURCE: (*12*)
SOURCE: (*13*)

Calculation

\$173,775 = 3,460 × 0.018 × 2,196 × 1.033 × 1.23

AID FOR FAMILIES WITH DEPENDENT CHILDREN

Savings = [Number AIS 3–5 Injured + Fatalities] × [Proportion of Population Receiving AFDC]
0.064
SOURCE: (*12*)

× [Average Annual Benefit] × [Administrative Adjustment] × [CPI]
$3,456

Calculation

$1,815,481 = 6,460(0.064)(3,456)(1.033)(1.23)

TITLE XX

Assume 2.4 percent of Medicaid + SSI + AFDC = $210,647
SOURCE: (*14*)

FOOD STAMPS

Estimation of the cost to the Food Stamp program requires an assumption about the number of households that would be receiving food stamps if the 55 mph speed limit were not in effect. This estimate is then multiplied times the average annual benefit to approximate the savings to the program.

Households currently using food stamps by AIS Level:		Injuries as a proportion of U.S. population:	Households avoiding the use of food stamps due to averted injuries to the 55 mph speed limit:
AIS-3 1,380,000	×	2,900/230 million	= 18
AIS-4 1,380,000	×	400/230 million	= 3
AIS-5 1,380,000	×	100/230 million	= 1
Fatals 1,380,000	×	3,000/230 million	= 18
			Total = 40

Savings = Households × Average Benefit × Administrative Adjustment × CPI = 40 × 1,246 × 1.03 × 1.23 = 63,143

PUBLIC EMPLOYEE COSTS

Public employee costs are derived from two different categories: workmen's compensation claims that result from automobile accidents and sick leave benefits to compensate workers when their injuries prevent them from working.

Workmen's Compensation

Federal savings = [Number of Injuries] × [Proportion of Severe Injuries Occurring to Federal Workers]
0.00172
× [Average Claim]
$3,098
SOURCE (*15*)

State and local savings = [Number of Injuries] × [Proportion Occurring to State and Local Employees]
0.00632
× [Average Claim]
$3,098
SOURCE: (*15*)

Federal savings = 3,460 × 0.00172 × $3,098 × 1.23 = 22,677

State and local savings = 3,460 × 0.00632 × $3,098 × 1.23 = 83,326

Total 106,003

Sick Leave

Injury Category	Number Injuries Averted	Estimated Days Sick	Total Days
AIS 1	39,000	2.51	97,890
AIS 2	8,500	13.75	116,875
AIS 3	2,900	37.19	107,851
AIS 4	400	125.0	50,000
AIS 5	160	125.0	20,000
			392,706

Federal

[Time Lost] × [Proportion of Federal Employees Injured]
0.00172
× [Average Daily Wage for Federal Employees] × [Administrative Cost] × [CPI] = $74,523
$78.00 1.15 1.23
SOURCE: (*15*)

State

[Time Lost] × [Proportion of State Employees Affected]
(see estimates above) 0.00632
× [Average Daily Wage for State Employees] × [Administrative Costs]
$58.50 × [CPI] = 205,373 1.15

Total Sick Leave Savings: Federal + State = Total
$74,523 + 205,373 = 279,896

Public Employee Savings: Workmen's Compensation + Sick Leave
$385,899 = 106,003 + 279,896

Savings in Indirect Costs

Police Savings (*16*)

Injury Category	Number Averted	Cost (1983 $)	Total Savings (000s)
AIS 1	39,000	47	$1,833
AIS 2	8,500	66	561
AIS 3	2,900	95	276
AIS 4	400	132	53
AIS 5	160	160	25
Fatal	3,000	157	471
			$3,219

Fire Department (*16*)

Fire Department Cost per Accident (AIS 4 − Fatal) = $44 per Accident × 3,560 = $160,000

Coroner/Medical Examiner (*16*)

Cost/Fatality = $191 × 3,000 = $573,000

Police + Fire + Coroner = Total Indirect Cost
$3,219,000 + 160,000 + 573,100 = $3,952,000.

REFERENCES

1. *Fatal Accident Reporting System 1981*. Figure 4. NHTSA, National Center for Statistics and Analysis, U.S. Department of Transportation, 1983.
2. *The Economic Cost to Society of Motor Vehicle Accidents*. NHTSA, U.S. Department of Transportation, 1983 (Hereafter referred to as NHTSA 1983).

3. NHTSA, U.S. Department of Transportation, 1983, p. A-6.
4. NHTSA, U.S. Department of Transportation, 1983, p. A-10.
5. NHTSA, U.S. Department of Transportation, 1983, p. A-7.
6. NHTSA, U.S. Department of Transportation, 1983, p. A-5.
7. NHTSA, U.S. Department of Transportation, 1983, p. A-9.
8. NHTSA, 1983, p. IV–13, Table IV.
9. *Statistical Abstract of the United States Table IV. 1982–83*. U.S. Bureau of the Census, pp. 334.
10. NHTSA, 1983, p. A-3.
11. NHTSA, 1983, p. A-9, 10.
12. *Statistical Abstract of the United States Table 554, 1982–83*. U.S. Bureau of the Census, pp. 340.
13. NHTSA, 1983, p. A-19.
14. NHTSA, 1983, p. A-26.
15. NHTSA, 1983, pp. A-37, A-56–67.
16. *Alternative Approaches to Accident Cost Concepts*. FHWA, U.S. Department of Transportation, 1984, pp. 22–36.

Appendix H
Further Research

During the course of this study the Committee for the Study of Benefits and Costs of the 55 mph National Maximum Speed Limit encountered a number of instances where necessary data were either inadequate or nonexistent. In other instances, the data were available, even if scattered, but had never been properly analyzed and it was not feasible to assemble and process such data for this study.

These gaps in our knowledge limit the ability to draw conclusions and generalizations about how decisions on speed limits will affect highway safety. In order to build a stronger foundation for policy analysis, the committee suggests that the responsible federal agencies begin to develop data bases and support the research discussed in the following sections.

SPEED DATA

The Federal Highway Administration currently collects speed data only on roads posted at 55 mph and publishes data only on the portions of those highways that have posted speeds exceeding 55, 60, or 65 mph, the mean speed, and the 85th percentile speed. Although such aggregate statistics may be useful for current monitoring purposes and for discerning broad trends, they are not always sufficiently detailed for the purposes of analysis. Moreover, current reporting requirements do not reveal the frequency of very high speed driving that is most threatening to highway safety. In addition to providing more detailed speed data on roads posted at 55 mph, a sample of vehicle speeds should be taken on highways posted below 55 mph. This resulting information will allow better estimation of the relationship between speed and safety, and will help determine if restricted speeds on well-patroled roads posted at 55 mph are causing motorists to drive faster on less safe systems where there is less likelihood that they will be apprehended.

RELATIONSHIPS BETWEEN SAFETY AND SPEED MEAN/VARIANCE

Previous research indicates that the variance in highway speeds is an important determinant of accident probability (*1*). Statistical analyses given in Chapter 3 and Appendix C highlight the importance of speed variance. The hypothesis offered in Appendix C, that only speed variance and not average speed matters to highway safety, is suggestive but lacks substantial empirical support at present. The implication of this hypothesis is that speed limits could be raised without adversely affecting safety as long as traffic continues to move at a uniform pace. This argument runs counter to other studies and analyses. As discussed in Chapter 3, crashes at higher impact speed have a much greater probability of resulting in fatality. Even if accident occurrence remained unchanged as a result of increased speed limits (assuming an unchanged variance) the accidents that did occur would of necessity be more severe because of higher average speeds.

In addition, whether speed limits could be raised without simultaneously increasing speed variance is uncertain. It is uncertain which policy measures, such as minimum speed limits, might be used to constrain the speed variance, nor is it known what their likely impacts would be. Further, there are physical and behavioral constraints on how effective such policies could be. For example, the wider variance in speeds occurs at intersections where the minimum speed is defined by the slower speed necessary for exiting and entering traffic. Higher average speeds would increase speed variance at these locations. Similarly, weather, congestion, and road geometry influences the degree to which more uniform speeds can be achieved on any given road segment at a particular time. Behavioral responses to variance-reduction policies are largely unknown. For example, many drivers, particularly the elderly, are more comfortable driving at slower speeds, and may continue to drive slower even if speed limits were raised. Persons 65 years of age and older are the fastest growing group in the nation. More research is needed to determine the relative importance of average speed and speed variance to highway safety, and further analysis is needed to determine whether average speed could be increased independently of speed variance, while accounting for the physical and behavioral factors involved.

SAFETY AND TRAVEL PURPOSES

The belief that recreational driving, which is more prone to random fluctuations than routine trips, is inherently more dangerous is alluded to in this study. Data are needed on the trip purposes of fatal and serious accident victims to test this hypothesis. Future evaluations of safety policy programs will require that some program of continuing trip exposure measurement be developed. As noted in Chapter 3, many changes in driving behavior may have occurred during the energy crisis. Measurement of these changes is impossible, except through indirect proxies. A trip exposure measurement program based on a sample of trips would greatly aid analysis of changes in driving behavior, policies aimed at altering driver behavior, and the effects of these changes on highway safety.

SAFETY AND MEASURES OF EXPOSURE

Virtually all measures of exposure focus on miles of travel (passenger miles, vehicle miles, or seat miles). Although this may be appropriate for intermodal comparisons, these measures do not precisely assess the safety impact of lengthening trip times for trips of a given distance. Although there is no well-established relationship between travel time and safety, the possibility that an offsetting negative effect exists cannot be completely discounted. This hypothesis has never been rigorously or empirically tested. Collecting data on how long accident victims have been driving would help identify this relationship.

SAFETY AND TRAFFIC DENSITY

The committee considered the possibility of raising speeds on low-density, rural Interstate highways, but rejected that approach partly because some low-density Interstates have a much poorer safety record than more heavily traveled Interstates. However, the data on the relationship between safety and traffic density are at the state level and do not relate strictly to traffic density on specific roads. Research on individual Interstate highway experiences, controlling for other factors, could help to better establish the relationship between safety and traffic density.

ENFORCEMENT ISSUES

More research is needed on the question of how motorist perceptions of enforcement affect compliance and how to best influence those perceptions. One important factor influencing deterrence is the role of traffic courts in meting out fines for speed-limit violators. As demonstrated in the analysis of state penalties and speed-limit violators in Chapter 9, it is not known whether the differences in state penalties among states are directly related to differences in traffic court stringency between states. Anecdotal evidence suggests that traffic courts vary in stringency as much within states as among them. A test of this hypothesis would require a considerably large sample and extensive data collection. It might be possible, however, to test this hypothesis on a subset of states.

Recent research in police patrol strategies (cited in chapter 9) has demonstrated that concentrated enforcement efforts that increase the visibility of enforcement do alter motorist behavior. More research is needed to determine how these concentrated efforts affect overall levels of speeding within a jurisdiction and how enforcement agencies can optimize the limited resources available for enforcing speed limits. The successful application of automated speed control devices in other nations, and field tests in the United States, suggest that these devices offer a cost-effective alternative to manned patrols (*2*); however, such devices raise complex constitutional and public acceptability issues. Further research is needed in both these areas.

REFERENCES

1. D. Solomon. *Accidents on Main Rural Highways Related to Speed, Driver, and Vehicle*. FHWA, U.S. Department of Transportation, 1964.
2. R. Blackburn and W. Glauz, *Pilot Tests of Automated Speed Enforcement Devices and Procedures*. NHTSA, U.S. Department of Transportation, February, 1984.

Appendix I
Detailed Tables

TABLE A–21 Fatality Rate (per 100 million vehicle miles), 1946–1983

Year	Rate	Percentage Change	Year	Rate	Percentage Change
1946	9.8	−12.65	1965	5.54	−1.60
1947	8.82	−10.00	1966	5.70	+2.89
1948	8.11	−8.05	1967	5.50	−3.51
1949	7.47	−7.89	1968	5.40	−1.82
1950	7.59	−1.61	1969	5.21	−3.52
1951	7.53	−0.79	1970	4.88	−6.33
1952	7.36	−2.26	1971	4.57	−6.35
1953	6.97	−5.30	1972	4.43	−3.06
1954	6.33	−9.18	1973	4.24	−4.29
1955	6.34	+0.16	1974	3.59	−15.33
1956	6.28	−0.95	1975	3.45	−3.90
1957	5.98	−4.78	1976	3.33	−3.48
1958	5.56	−7.02	1977	3.35	+0.60
1959	5.41	−2.70	1978	3.39	+1.19
1960	5.31	−1.85	1979	3.50	+3.20
1961	5.16	−2.82	1980	3.50	0.00
1962	5.32	+3.10	1981	3.32	−5.14
1963	5.41	+1.69	1982	2.93	−11.75
1964	5.63	+4.07	1983	2.76	−5.80

SOURCE: National Safety Council. *Accident Facts, 1982*. Fatality rate estimate for 1983 provided by the National Safety Council.

TABLE A–22 Average Speeds on Selected Highways, 1973–1983

Highway System	1973 Average Speed (mph)	1974 Average Speed (mph)	Change in Speed 1973–74 (mph)	1983 Average Speed (mph)	Change in Speed 1974–83 (mph)	Retained Benefits[a] (%)
Rural Interstate	65.0	57.6	−7.4	59.1	+1.5	79.7
Urban Interstate	57.0	53.1	−3.9	56.8	+3.7	5.1
Rural primary	57.1	53.5	−3.6	54.6	+1.1	69.4

SOURCE: *Highway Statistics*, Table VS–1 and 2, 1974, FHWA, U.S. Department of Transportation. Estimates for 1983 provided by FHWA.

[a]The retained benefits of the speed limit is defined as the percentage of the speed reduction retained today relative to 1974 or

$$\left[\frac{(\text{speed 1973} - \text{speed 1983})}{(\text{speed 1973} - \text{speed 1974})}\right] \times 100$$

TABLE A–23 Estimated Effect on Fatalities of Demographic Shifts, 1982 versus 1973

Age Groups	Licensed Drivers 1973[a] (%)	Licensed Drivers 1982[b] (%)	Driver Involvements in Fatal Accidents 1982[c]	Estimated Driver Involvements in Fatal Accidents[d] 1973	Effect of Demographic Shifts[e]
16–24	22.2	20.2	19,603	21,544	−1941
25–44	39.5	42.9	23,138	21,305	+1833
45–64	29.2	25.8	9,074	10,269	−1195
65+	9.1	11.1	3,954	3,242	712
Total			55,769	56,360	−591

[a]*Highway Statistics, 1973*. Table DL–21B. U.S. Department of Transportation.
[b]*Highway Statistics, 1982*. Table DL–21B. U.S. Department of Transportation.
[c]*Fatal Accident Reporting System*. Table 62. U.S. Department of Transportation, 1982. *Note*: 1,145 drivers of unknown age were distributed across age groups based on the proportion of known driver involvements by age.
[d]This estimate is derived by multiplying the ratio of percentage of licensed drivers in each age group for 1972 by the percentage of licensed drivers in 1982 times the number known fatal involvements for each age group in 1982.
[e]The effect of demographic shifts is estimated by subtracting the known estimated driver involvements in each age group from the known driver involvements in each age group.

TABLE A–24 Estimated Distribution of Nonfatal Injuries by Maximum AIS by Highway Class 1982[a]

Highway System	AIS 1 (%)	AIS 2 (%)	AIS 3 (%)	AIS 4 (%)	AIS 5 (%)	Total (%)
Urban Interstates and freeways	82.65	12.84	4.01	0.47	0.02	100
Rural Interstates	69.02	21.79	8.32	0.66	0.21	100
Rural arterials	77.06	16.40	5.64	0.60	0.30	100
Rural secondary	79.01	13.37	6.49	0.71	0.42	100

[a]Estimates provided by the National Center for Statistics and Analysis, NHTSA, U.S. Department of Transportation, based on 1982 National Accident Sampling System data.

TABLE A–25 Estimated Reductions in Injuries by AIS Level and Highway Type[a], 1973–1974

	AIS 1			AIS 2			AIS 3			AIS 4			AIS 5		
Highway Classification[b]	1973	1974	Change	1973	1974	Change	1973	1974	Change	1973	1974	Change	1973	1974	Change
Urban Interstates and freeways	121,539	101,282	20,957	18,879	15,733	3,146	5,896	4,913	983	691	576	115	30	25	5
Rural Interstates	60,351	41,817	18,534	19,053	13,208	5,854	7,275	5,029	2,246	577	399	178	184	122	62
Rural other federal-aid primary	235,016	201,444	33,572	50,016	42,872	7,144	17,201	14,744	2,457	1,830	1,568	262	915	784	131
Rural federal-aid state	140,028	133,877	6,151	23,695	22,655	1,040	11,502	10,997	505	1,258	1,203	55	744	711	33
Total reductions			79,214			17,184			6,191			610			231

[a]These calculations assume 1973 and 1974 total injuries are distributed according to the injury distributions given in Table A–24.
[b]Estimates of total nonfatal injuries by highway type from *Fatal and Injury Accident Rates on Federal-Aid and Other Highway Systems*, 1975. FHWA, U.S. Department of Transportation, Tables IT–1, IT–31, IT–3, IT–5. See *Revised Total* columns for 1973 and 1974 estimates.

TABLE A–26 Petrolum Imports and Domestic Petroleum Production, Selected Years 1960–1980 (000s BPD)

Year	Domestic Production Crude Oil	Total	Imports[a] Refined Petroleum Products	Crude Oil	Domestic Petroleum Consumption[b]	Imports as a Percentage of Domestic Consumption
1960	7,989	1,820	–[c]	–[c]	9,825	18.5
1970	11,312	3,419	–[c]	–[c]	14,696	23.2
1973	9,208	6,256	3,012	3,244	17,308	36.1
1974	8,774	6,112	2,635	3,477	16,653	36.7
1975	8,375	6,056	1,951	4,105	16,322	37.1
1976	8,132	7,313	2,026	5,287	17,461	41.9
1977	8,245	8,807	2,193	6,615	18,431	47.8
1978	8,707	8,363	2,008	6,356	18,847	44.4
1979	8,552	8,456	1,937	6,519	18,513	45.7
1980	8,597	6,909	1,646	5,263	17,056	40.5
1981	8,572	5,996	1,599	4,396	16,058	37.3
1982	8,649	5,113	1,625	3,488	15,296	33.4
1983[d]	8,664	4,935	1,629	3,306	15,004	32.9

SOURCE: H.R. Millie, ed. *Minerals and Materials: A Monthly Survey*. Bureau of Mines, U.S. Department of Interior, 1976, p. 7. *Monthly Energy Review*. Energy Information Administration, U.S. Department of Energy, Nov. 1983, pp. 32–3.

[a]Excludes imports for strategic petroleum reserve (SPR), which began in October 1977.

[b]Domestic pretroleum consumption is defined as refined petroleum products supplied. Refined petroleum products are those obtained from the processing of crude oil, unfinished oils, natural gas liquids, and other miscellaneous hydrocarbon compounds.

[c]Data not available.

[d]Through September 1983.

TABLE A–27 Effect of Weight on Passenger Car Fuel Consumption by Inertia Weight Class[a]

Inertia Weight Class (lb)	Fuel Consumption Rate at 55 mph (gal/mile)	Miles per Gallon at 55 mph	Difference in Fuel Consumption Rate Between 55 mph and Indicated Speed					Sample Size
			45 mph	50 mph	60 mph	65 mph	70mph	
2,000	0.0283	35.3	−11.5	−6.2	7.0	15.0	23.7	69
2,250	0.0289	34.6	−11.5	−6.6	8.5	18.7	30.8	148
2,500	0.0328	30.5	−11.5	−6.6	6.8	15.4	25.5	131
2,750	0.0356	28.1	−8.9	−5.3	8.5	18.7	30.8	74
3,000	0.0406	24.6	−9.1	−5.1	6.1	13.1	21.2	252
3,500	0.0456	21.9	−5.4	−3.3	5.3	13.2	24.3	368
4,000	0.0506	19.8	−8.1	−4.5	5.4	11.7	18.9	323
4,500	0.0539	18.6	−7.0	−4.0	4.9	10.0	17.8	279
5,000	0.0580	17.2	−7.0	−3.7	4.3	9.1	14.4	71

SOURCE: R.L. Mason and R.W. Zub. *Highway Fuel Economy Study*, Transportation System Center, U.S. Department of Transportation, 1981.

[a]Diesel-powered vehicles not included.

TABLE A–28 Estimated Highway Fuel Consumption Rates for Passenger Cars

Model Year	Percent of Travel Mileage by Model Year	Fuel Consumption Rate for 1972 Fleet (gal/mile)	Fuel Consumption Rate for 1979 Fleet (gal/mile)
1980	0.058	0.0426	0.0402
1979	0.129	0.0469	0.0442
1978	0.125	0.0485	0.0458
1977	0.115	0.0525	0.0495
1976	0.098	0.0570	0.0537
1975	0.076	0.0593	0.0559
1974	0.073	0.0580	0.0547
1973	0.084	0.0580	0.0547
1972	0.066	0.0576	0.0543
1971	0.050	0.0568	0.0536
1970	0.033	0.0566	0.0534
1969	0.029	0.0576	0.0543
1968	0.021	0.0565	0.0532
Pre–1968	0.043	0.0557	0.0525
Average		0.0536	0.0500

SOURCE: R.L. Mason and R.W. Zub. *Highway Fuel Economy Study*. Transportation Systems Center, U.S. Department of Transportation, 1981.

TABLE A–29 Estimated Fuel Consumption and Fuel Economy Rates of 1983 Passenger Automobile Fleet for Before and After the 55 mph Speed Limit[a]

		Fuel Consumption[b]			
Model Year	Percent of Annual Mileage by Model Year	Pre–1974 Gal/Mile	Post–1974 MPG	Gal/Mile	MPG
1983	0.058	0.0331	30.2	0.0306	32.7
1982	0.129	0.0363	27.5	0.0337	29.7
1981	0.125	0.0396	25.2	0.0369	27.1
1980	0.115	0.0426	23.5	0.0402	24.9
1979	0.098	0.0469	21.3	0.0442	22.6
1978	0.076	0.0485	20.6	0.0458	21.8
1977	0.073	0.0525	19.0	0.0495	20.2
1976	0.084	0.0570	17.5	0.0537	18.6
1975	0.066	0.0593	16.9	0.0559	17.9
1974	0.050	0.0580	17.2	0.0547	18.3
1973	0.033	0.0580	17.2	0.0547	18.3
1972	0.029	0.0576	17.4	0.0543	18.4
1971	0.021	0.0568	17.6	0.0536	18.6
(1963–1970)	0.043	0.0566	17.7	0.0534	18.7
Average		0.0476	21.0	0.0445	22.5

[a]Estimates based on Table A–28.
[b]Fuel consumption estimates depend on the speed distribution in the pre- and post-55 mph speed limit time periods.

TABLE A–30 Crude Oil and Petroleum Product Imports (000s BPD)

Year	Total OPEC	Total Arab OPEC	Non-OPEC	Total Imports	Arab OPEC as a Percent of Total Imports
1973	2,993	915	3,263	6,256	14.6
1974	3,280	752	2,832	6,112	12.3
1975	3,601	1,383	2,454	6,055	22.8
1976	5,066	2,424	2,247	7,303	33.2
1977	6,193	3,185	2,614	8,807	36.2
1978	5,751	2,963	2,613	8,364	35.4
1979	5,637	3,056	2,819	8,456	36.1
1980	4,300	2,551	2,609	6,909	36.9
1981	3,323	1,848	2,672	5,995	30.8
1982	2,146	854	2,968	5,114	16.7
1983	1,676	483	3,120	4,796	10.1

SOURCE: *Monthly Energy Review*. Energy Information Administration, U.S. Department of Energy, Nov. 1983.

TABLE A–31 Estimated Additional Travel Time by Length of Trip for Trips Greater Than 10 Miles

(1) Trip Length (miles)	(2) Additional Minutes of Driving[a]	(3) Percentage of Total Vehicle Miles of Travel[b]	(4) Estimated Personal Travel on Highways Posted at 55 mph	(5) Estimated Occupancy Rate[c]	(6) Estimated Passenger Travel[d]	(7) Percent
11–20	0.7–1.25	36.7	210	1.68	353	31.8
21–30	1.26–1.9	19.2	110	1.74	191	17.2
31–50	2.0–3.1	16.3	95	2.02	192	17.3
51–100	3.2–6.2	12.2	70	2.34	164	14.7
More than 100	More than 6	15.6	90	2.34	211	19.0

[a]See Table A–32 for comparisons of 1973 and 1982 average speeds.
[b]Total passenger travel distributed for trips of greater than 10 miles as found in "Purposes of Vehicle Trips and Travel," *1977 Nationwide Personal Transportation Study*, Report 3, Table A30, 1980
[c]See "Vehicle Occupancy," *1977 Nationwide Personal Transportation Study*, Report 6, Table 5, 1981.
[d]Column 4 multiplied by column 5.

TABLE A–32 Estimated Average Speeds for Passenger Vehicles, 1973 and 1982

(1) Highway Class	(2) Proportion of Travel of 55 mph Highways[a]	(3) Average Speed 1973[b]	(4) Weighted Speed 1973[c]	(5) Average Speed 1982[b]	(6) Weighted Speed 1982[d]
Rural Interstate	0.1843	66.6	12.27	59.0	10.87
Other rural primary	0.3157	58.2	18.37	54.6	17.24
Rural secondary	0.1533	53.2	8.15	51.6	7.9
Urban Interstate and freeways	0.3466	58.0	20.10	56.3	19.5
Total	0.999		58.9		55.5

[a]Estimates provided by the Federal Highway Administration.
[b]*Highway Statistics*, Tables VS–1 and VS–2, 1973 and 1982, FHWA, U.S. Department of Transportation.
[c]Column 2 multiplied by column 3.
[d]Column 2 multiplied by column 5.

TABLE A–33 Additional Vehicle Hours of Travel

Highway System	(1) 1982 VMT (millions)[a]	(2) 1982 Average Speed[b]	(3) 1973 Average Speed[c]	(4) Additional Hours (billions)[d]
Rural Interstates	140,778	59.0	65.0	0.22025
Rural arterials	223,903	54.5	57.1	0.20220
Rural collectors	106,950	51.6	52.6	0.0394
Urban Interstates and freeways	231,972	56.3	57.0	0.05059
Total				0.51244

[a]Travel on all highways posted at 55 mph provided by FHWA. Travel on urban arterials (42,243 million VMT) excluded because these highways are assumed to have been posted at 55 mph before 1973.
[b]*Highway Statistics, 1982*. Table VS–1. Urban Interstates and urban freeways assumed to have the same speed in 1982 and 1973; 1982 speed on urban freeways was actually 0.5 mph slower, but comparable speeds for 1973 are not available classified by urban freeway.
[c]*Highway Statistics, 1973*. Table VS–1. Rural arterial highways are assumed to be the same as the rural primary classification used in 1973. Rural collectors are assumed to be the same as the rural secondary classification used in 1973.
[d]Estimates in column 4 derived by the following formula for each highway system:

$$\left[\frac{(1982\ \text{VMT})}{(1982\ \text{Average Speed})} - \frac{(1982\ \text{VMT})}{(1973\ \text{Average Speed})}\right]$$

TABLE A–34 Additional Passenger Hours of Travel

	Additional Vehicle Hours (000s)[a]	Occupancy Rates[b]	Additional Passenger Hours (000s)
Cars and light trucks	418,407	1.93	807,526
Commercial trucks	92,034	2.0	184,068
Buses	1,998	23.0	45,966
Total	512,440		1,037,560

[a]Distribution of vehicle hours based on vehicle miles traveled given in Table 22, Chapter 7. See Table A–33 for estimate of total vehicle hours of additional travel.
[b]Passenger car occupancy rates are derived from the "Vehicle Occupancy," *1977 Nationwide Personal Transportation Study*, Report 6, Table 5 for trips more than 11 miles. Truck occupancy rates estimated by TRB. Bus occupancy rates estimated based on trends provided in *Transportation Facts and Trends*, Transportation Association of America, 1980, Table A–8.

Study Committee Biographical Information

ALAN A. ALTSHULER, *Chairman*, is a political scientist and is currently Dean of the Graduate School of Public Administration, New York University. He served as Secretary of Transportation and Construction, Commonwealth of Massachusetts, 1971-1975; Director, Boston Transportation Planning Rev., 1970-1971; and chaired the Governor of Massachusetts' Task Force on Transportation, 1969-1970. Dr. Altshuler was elected to membership in the National Academy of Public Administration in 1975. He received his B.A. from Cornell University and his M.A. and Ph.D. from the University of Chicago. He has held the following positions: Assistant Professor, Cornell University, 1962-1966; Associate Professor, Massachusetts Institute of Technology, 1966-1969; Professor, Massachusetts Institute of Technology, 1969-1971, 1975-1983. He served as head of the MIT Department of Political Science from 1977-1982. Coauthor of *The Future of the Automobile*, 1984, Dr. Altshuler's publications include the following books: *The Urban Transportation System: Politics and Policy Innovation*; *The City Planning Process: A Political Analysis*; *Community Control: The Black Demand for Participation in Large American Cities*; and *The Politics of the Federal Bureaucracy*.

HARRY T. ADAIR, a law enforcement administrator, is the Assistant Commissioner/Staff of the California Highway Patrol. Having joined the California Highway Patrol in 1963, Mr. Adair commanded various field operations throughout the state of California. As Assistant Commissioner/Staff, he directs all planning, analysis, training, administrative, and specialized enforcement functions of the Highway Patrol.

He holds a B.A. in economics and a masters degree in public service from California State University at Los Angeles. Mr. Adair is a member of the California Peace Officers Association, the International Association of Chiefs of Police, and the California Association of Police Planning and Research Officers.

JOHN R. BORCHERT is an educator and currently Regents Professor of Geography, University of Minnesota. Elected to membership in the National Academy of Sciences in 1976, Dr. Borchert has served on numerous National Research Council committees since 1955 and is currently a member of the TRB Executive Committee. In addition, he has served as Urban Research Director, Upper Midwest Economic Study, 1961-1963; Project Director, Minnesota Highway Research Project, 1958-1961; and Director, Center for Urban and Regional Affairs, University of Minnesota, 1968-1977. Dr. Borchert earned his B.A. from DePauw University and his M.S. and Ph.D. from the University of Wisconsin. He served as president of the Association of American Geographers, 1968-1969, and was appointed Regents Professor of Geography at the University of Minnesota in 1982.

JOSEPH M. CLAPP is Senior Vice-President and Director of Roadway Express, Incorporated, Akron, Ohio, and is chairman of the TRB Executive Committee. A graduate of the Traffic and Transportation Program of the University of North Carolina in 1958, Mr. Clapp served as Director, Safety and Personnel, Ryder Truck Lines, Incorporated, and Assistant to the President, McCormack Trucking Company, Incorporated, before joining Roadway Express. Mr. Clapp is former chairman of the North Carolina Motor Carrier Association Council of Safety and Personnel Supervisors; a member of the TRB Committee on Freight Transport Service Quality; Chairman of the Regular Common Carrier Conference; member, American Trucking Association, Safety and Engineering Committee; and the Association of Transportation Practitioners.

WALTER DALE COMPTON, a physicist, is Vice President of Research, Ford Motor Company. A graduate of Wabash College of Physics with a B.A. Degree, Dr. Compton received his M.S. from the University of Oklahoma and his Ph.D. from the University of Illinois. He is a member of the National Academy of Engineering and a Fellow of the American Physics Society. His research includes solid state physics, radiation effects in solids, color centers in insulating crystals, luminescence, and metal semiconductor junction. Dr. Compton has held positions at the U.S. Naval Weapons Center China Lake, 1951-1952, U.S. Naval Research

Laboratory 1955-1961, taught physics at the University of Illinois, Urbana, 1961-1970, and served as Director, Coordinated Science Laboratory, University of Illinois, 1965-1970. He is a recipient of the U.S. Naval Research Laboratory Award.

R ADAMS COWLEY is a physician and is currently Director, Maryland Institute of Emergency Medical Services Systems. Dr. Cowley received his MD from the University of Maryland and has been a Professor of Thoracic and Cardiovascular Surgery at the University of Maryland since 1961. He was a member of the National Highway Safety Advisory Committee from 1978 to 1981 and currently serves on the Maryland Highway Safety Coordinating Committee. Dr. Cowley is a member of numerous professional associations, including: the American Association of Automotive Medicine, American Trauma Society, American Association of Surgery Trauma, Society of Critical Care Medicine, and the American College of Emergency Room Physicians.

BENJAMIN O. DAVIS, Lt. General (Retired), United States Air Force. After a distinguished military career including combat service as a fighter group commander in World War II, jet fighter Wing Commander in Korea, and 13th Air Force Commander in the Phillipines, General Davis served as Assistant Secretary of Transportation for Environment, Safety, and Consumer Affairs, 1971-1975. He is a graduate of the U.S. Military Academy with a B.S. degree, and the Air War College, Maxwell Air Force Base, Alabama. Among the many honors General Davis has received are the Distinguished Service Medal (Army, Air Force); Legion of Merit, Silver Star, D.F.C., Air Medal with eight oak leaf clusters (U.S.); Croix de Guerre with palm (France); Star of Africa, Republic of China Cloud and Banner Medal with Oak Leaf Cluster; Phillipine Legion of Honor; and the Republic of Vietnam Campaign Medal.

JOHN J. FEARNSIDES is Technical Director, Air Transportation Systems Development Division at the MITRE Corporation. He received his B.S. and M.S. from Drexel University and his Ph.D. from the University of Maryland. Dr. Fearnsides served as Director, Economic and Management Sciences Division, Analytical Sciences Corporation, 1979-1980. From 1972 to 1979 he held a number of positions at the U.S. Department of Transportation, including Acting Assistant Secretary for Policy and International Affairs; Deputy Under Secretary/Chief Scientist; Acting Administrator, Research and Special Programs Administration; Executive Assistant to the Deputy Secretary; Acting Director, Materials Transportation Bureau; Chief, Research and Development, Policy Division; Special Assistant to the Assistant Secretary for Systems Development and Technology; and Manager, Advanced Research Program. Dr. Fearnsides has published numerous articles in aerospace engineering. He is a member of the Institute of Electrical and Electronic Engineers and Sigma Xi.

PAUL H. FOWLER is a registered traffic engineer in the state of California. Since 1965 he has served as Traffic Engineer, Auto Club of Southern California, Los Angeles. He was Assistant City Traffic Engineer for the city of San Diego from 1952 to 1955 and later served as County Traffic Engineer from 1955 to 1965. Mr. Fowler is a graduate of the University of the Pacific, Stockton, California, with a B.S. in civil engineering, and the Yale University School of Traffic Engineering. He is a Director of the Institute of Transportation Engineers and a member of its National Policy, Legislation, and Safety Committees, and the following TRB Committees: Traffic Control Devices, Chair; Motorist Services Committee, past chair; Free Operations Committee; panelist for NCHRP report, "Motorist Response to Freeway Directional Signing." Mr. Fowler is also a member of the National Committee on Uniform Traffic Control Devices, American Public Works Association, AAA Advisory Committee on Traffic and Safety, California Traffic Control Devices Committee, and the governor's Traffic Safety Conference (past chair). He is the author of several articles and booklets on traffic engineering and is a frequent Instructor for the Institute of Transportation Studies, University of California.

TREVOR O. JONES is Vice President and General Manager, Transportation, Electronics Group, TRW, Incorporated. Mr. Jones is past Vice Chairman of the Department of Transportation National Motor Vehicle Safety Advisory Council and past Chairman of the National Highway Safety Advisory Committee. He received the Higher National Certificate in Electrical Engineering from Aston College, Birmingham, England, and the Ordinary National Certificate in Mechanical Engineering, Liverpool Technical College, Liverpool, England. Mr. Jones was Director, General Motors Proving Grounds, 1974-1978, and Vice President, Engineering, Automotive Worldwide TRW, Incorporated, 1978-1981. In 1978 he received an award from the U.S. Department of Transportation for excellence in engineering. Elected to the National Academy of Engineering in 1982, Mr. Jones is a registered professional engineer in Wisconsin and a chartered engineer, United Kingdom, Fellow British Institute of Electrical Engineers. He is a Fellow of the Society of Automotive Engineers and received the Arch T. Colman paper award in 1974 and 1975, and the Vincent Bendix Automotive Electronics Award 1976.

CHARLES A. LAVE is a Professor of Economics, University of California, Irvine, where he was Chairman of the Economics Department, 1978-1983, and Chair of the Faculty of Social Sciences, 1978-1984. Dr. Lave received his B.A. from Reed College and his Ph.D. from Stanford University. He has published extensively on the subjects of transportation, energy, and the impact of the automobile on society. Past positions have included: Visiting Scholar at Harvard, Stanford, MIT, and Hampshire College; Senior Research Associate at Charles River Associates; and Consultant at Oak Ridge National Laboratory, 1975-1983. He has served as a member of the Board of Trustees of Reed College; Controller of Reflection Educational Products; and is now Chairman of the Board of the Irvine Campus Housing Authority. He is a member of the American Economic Association, the Econometric Society, and the Transportation Research Board; and is a Fellow of the Society for Applied Anthropology.

DARRELL V MANNING is Director of the Idaho Department of Transportation and Chairman of Engineering Operations. He is a past chairman of the TRB Executive Committee and has served as Vice President of the American Association of State Highway and Transportation Officials. Mr. Manning received his B.S. from Utah State University and has done graduate studies in public administration and management. He served five terms in the Idaho State Legislature, two as minority floor leader, 1960-1968 and three in the state senate. He served on the State Government Reorganization Commission, Idaho Constitutional Review Commission, Federal Land Law Review Commission, and the Legislative Council. He also served on the Board of Directors, International Northwest Aviation Council, and served as Regional Vice President of the Board of Directors, National Association of State Aviation Officials.

H. LAURENCE ROSS, an educator, is Chairman, Department of Sociology, University of New Mexico. He received his B.A. from Swarthmore College and his M.A. and Ph.D. from Harvard University. Dr. Ross has held the following positions: Instructor, Sociology, Northwestern University, 1959-1960; Assistant and Associate Professor, New York University, 1960-1967; Professor, Sociology and Law, University of Denver, 1967-1980; and Professor of Sociology and Law, State

University of New York, Buffalo, 1980-1983. He has served on the following Transportation Research Board Committees since 1970: Traffic Records; Operator Regulations; and Alcohol, Other Drugs, and Transportation. He has served on the Committee on Alcohol and Other Drugs, National Safety Council since 1972. Dr. Ross is a member of the American Sociological Association, the Society for the Study of Social Problems, the Law and Sociology Association, and has done research on the following topics: sociology of law, evaluation of legal impact, and the study of law in action. His most recent book is titled *Deterring the Drinking Driver: Legal Policy and Social Control.*

PETER J. SAFAR, a physician, is an anesthesiologist and critical care medicine physician. He is a distinguished Professor of Resuscitation Medicine and Founder and Director of the Resuscitation Research Center at the University of Pittsburgh. Dr. Safar received his M.D. from the University of Vienna in 1948 after which he served as Resident Pathologist and Surgeon, University of Vienna, 1948-1949; Fellow Surgeon, Yale University, 1949-1950; Resident Anesthesiologist, University of Pennsylvania, 1950-1952; Chief Anesthesiologist, Baltimore City Hospital, 1955-1961; Professor Anesthesiology/Critical Care Medicine and chairman of the department, University of Pittsburgh, 1961-1978. Dr. Safar has done research on resuscitation neuroscience, emergency medical services, and intensive care, and has authored and coauthored many publications. He was a member of the Committees on Resuscitation and Emergency Medical Services, the National Research Council; and the Interagency White House Committee on Emergency Medical Services. He was founding member and President of the Society of Critical Care Medicine. He is also President of the World Association for Emergency and Disaster Medicine.

WAYNE W. SORENSON, an insurance executive, is Vice President of Research, State Farm Mutual Automobile Insurance Company. He received his B.A. in business administration from the University of Minnesota and his M.A. and Ph.D. in industrial psychology from the University of Minnesota. A certified psychologist in the state of Illinois, Dr. Sorenson is a member of the American Psychological Association, The American Psychological Association Insurance Trust, the Highway Loss Data Institute Board of Directors, and the All-Industry Research Advisory Council. These latter two organizations have produced extensive insurance-related research on the subject of losses associated with automobile crashes. Dr. Sorenson has published articles in industrial psychology and automobile crashworthiness, one of which received the Arch T. Coleman award from the Society of Automotive Engineers.

EVELYN TEEGEN is currently chairman of the Subcommittee on Highway Environment and 55 mph of the National Highway Safety Advisory Committee of the National Highway Traffic Safety Administration. She holds a B.S. from Iowa State University and is Republican National Committeewoman, Independent Republicans of Minnesota. She is a member of advisory committees for candidates for national, state, and local elections and numerous community and civic organizations.

PATRICIA F. WALLER, an educator, is Associate Director for Driver Studies, in the University of North Carolina Highway Safety Research Center and serves as Research Professor in the Department of Health Policy and Administration in the University of North Carolina School of Public Health. She received A.B. and M.S. degrees from the University of Miami and her Ph.D. from the University of North Carolina. She is past president of the American Association for Automotive Medicine and has served on the National Highway Safety Advisory Committee. She has been active in the Transportation Research Board where she has chaired a variety of activities. Dr. Waller is also an Associate Editor of *Accident Analysis and Prevention* and serves on the editorial board of the *Journal of Safety Research*. Dr. Waller has authored numerous publications on the behavioral aspects of highway safety, methods of determining high risk drivers, driver education and training, stress and driving, and evaluation of highway safety programs.

JOHN J. WIORKOWSKI, an educator, is Professor of Statistics, Graduate Program in Mathematical Science, University of Texas, Dallas. He received his B.S., M.S., and Ph.D. in statistics from the University of Chicago. Dr. Wiorkowski has held the following positions: Assistant Professor of Statistics, Graduate Program of Health Care Administration, U.S. Army-Baylor University, 1968-1971; Research Associate, University of Chicago, 1972-1973; and Assistant Professor, Pennsylvania State University, 1973-1975. He has published numerous articles in applied statistics, specifically biostatistics, linear models, time series analysis, and genetic statistics. Dr. Wiorkowski is a member of Sigma Xi, the American Statistical Association, and the Institute of Mathematic Statistics.

MARCUS L. YANCEY is Deputy Engineer-Director of Administration, Texas State Department of Highways and Public Transportation. A graduate of the University of Texas with a B.S. in mechanical engineering, Mr. Yancey served as Texas Highway Department Bridge Design Engineer from 1957 to 1967. He is chairman of the American Association of State Highway and Transportation Officials (AASHTO) Highway Subcommittee on Design and a former vice chairman of the AASHTO Standing Committee on Planning. Mr. Yancey is also chairman of the Subcommittee on Legal Affairs and Highway Transport, National Conference of State Railway Officials.